ALL THUMBS
Guide to
Fixing Furniture

Other All Thumbs Guides
Car Care
Compact Disc Players
Home Computers
Home Energy Savings
Home Plumbing
Home Security
Home Wiring
Painting, Wallpapering, and Stenciling
Repairing Major Home Appliances
VCRs

ALL THUMBS
Guide to Fixing Furniture

Robert W. Wood
Illustrations by Steve Hoeft

TAB Books
Division of McGraw-Hill, Inc.
Blue Ridge Summit, PA 17294-0850

FIRST EDITION
FIRST PRINTING

© 1994 by **TAB Books.**
TAB Books is a division of McGraw-Hill, Inc.

Printed in the United States of America. All rights reserved. The publisher takes no responsibility for the use of any of the materials or methods described in this book, nor for the products thereof.

Library of Congress Cataloging-in-Publication Data

Wood, Robert W. (Robert Wynn)
   All thumbs guide to fixing furniture / by Robert W. Wood.
     p. cm.
   Includes index.
   ISBN 0-8306-4433-4
   1. Furniture—Repairing—Amateurs' manuals. 2. Furniture finishing—Amateurs' manuals. I. Title.
TT199.W66 1993
684.1'044—dc20   93-8044
                                            CIP

Acquisitions editor: Stacy Varavvas Pomeroy
Editorial team: Susan Wahlman, Editor
                Joanne Slike, Executive Editor
                Stacey Spurlock, Indexer
Production team: Katherine G. Brown, Director
                  Susan E. Hansford, Typesetting
                  Wanda S. Ditch, Layout
                  Nancy Mickley, Proofreading
Design team: Jaclyn J. Boone, Designer
             Brian Allison, Associate Designer
Cover design: Lori E. Schlosser
Cover illustration: Denny Bond, East Petersburg, Pa.   ATS
Cartoon caricature: Michael Malle, Pittsburgh, Pa.   4414

The All Thumbs Guarantee

TAB Books/McGraw-Hill guarantees that you will be able to follow every step of each project in this book, from beginning to end, or you will receive your money back. If you are unable to follow the All Thumbs steps, return this book, your store receipt, and a brief explanation to:

All Thumbs
P.O. Box 581
Blue Ridge Summit, PA 17214-9998

About the Binding

This and every All Thumbs book has a special lay-flat binding. To take full advantage of this binding, open the book to any page and run your finger along the spine, pressing down as you do so; the book will stay open at the page you've selected.

The lay-flat binding is designed to withstand constant use. Unlike regular book bindings, the spine will not weaken or crack when you press down on the spine to keep the book open.

Contents

Preface *ix*
Introduction *xi*
1 Safe Work Habits *1*
2 Tools for the Job *8*
3 Work Techniques *16*
4 Repairing Chairs *38*
5 Repairing Tables *58*
6 Drawers & Doors *77*
7 Furniture Touch-up *92*
8 Preparing the Surface for the Finish *106*
9 Finishing *116*
Glossary *123*
Index *127*

Preface

A collection of books about home repairs and improvements, the All Thumbs series was created not for the skilled jack-of-all-trades, but for the average person. If your familiarity with the various systems in the home is minimal, or if your budget isn't keeping pace with today's climbing costs, this series is tailor-made for you.

Several different professions are involved in the construction and maintenance of even the smallest home. The skills required seldom overlap. Carpenters build the framework, plumbers install the water system, and electricians complete the wiring. Indeed, few people can do it all. These crafts often require years of apprenticeship to master. The professional works quickly and efficiently and depends on a large volume of work to survive. Because service calls are time-consuming, often requiring more travel time than actual labor, they can be expensive. The All Thumbs series saves you time and money by showing you how to make most common repairs yourself.

The All-Thumbs series covers topics such as home wiring; plumbing; painting and wallpapering; repairing major home appliances; car care; home security, and this book, *Fixing Furniture*, to name a few. Each book breaks down many repairs or home improvements into easy-to-follow, well-illustrated steps within the ability of nearly anyone.

Introduction

It is a rare person indeed who hasn't been annoyed by a wobbly table or chair or fought with a sticking cabinet drawer. Time and frequent opening and closing can cause cabinet doors to stick or sag. Most of the problems can be solved with a few basic tools, wood glue, and only a little effort.

When furniture becomes scratched or dull, it often can be rejuvenated without stripping and refinishing. Refinishing a piece of furniture can be rewarding, but it also can be time-consuming and hazardous because of the toxic or flammable chemicals required. Any hazard can be overcome by a few basic rules, such as following the manufacturer's instructions and warnings and working in a well-ventilated room without any pilot lights or better yet, outdoors in the shade.

The first three chapters of this book cover safe work habits, tools, and how to use them. The next three chapters explain how to repair the most common problems with chairs, tables, and cabinet drawers and doors. The last three chapters show you how to repair surfaces, prepare the surfaces for finishing, and apply the finish.

When repairing furniture or doing any other home project, understand what you are going to do before you do it, don't be in a hurry, and be careful with power tools.

CHAPTER ONE

Safe Work Habits

Most accidents can be prevented by using common sense; however, mishaps do occur, even to the very careful. Mistakes happen. Paints and solvents spill, and power tools malfunction. Have a first aid kit in your home and know how to use it. Try to work in an area that is clear of debris, and don't hurry. The following steps illustrate safe work habits.

2 Fixing Furniture

Step 1-1. Be prepared.
Post emergency and poison-control phone numbers near the telephone, and instruct family members to stay calm and call 911 if an emergency occurs.

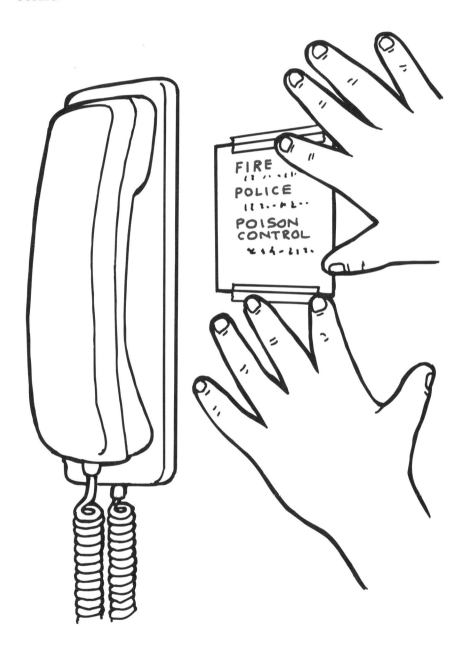

Step 1-2. Be careful.
Keep tools and chemicals out of the reach of children. Store chemicals away from heat. Thoroughly dry chemical-soaked rags outside in the open air.

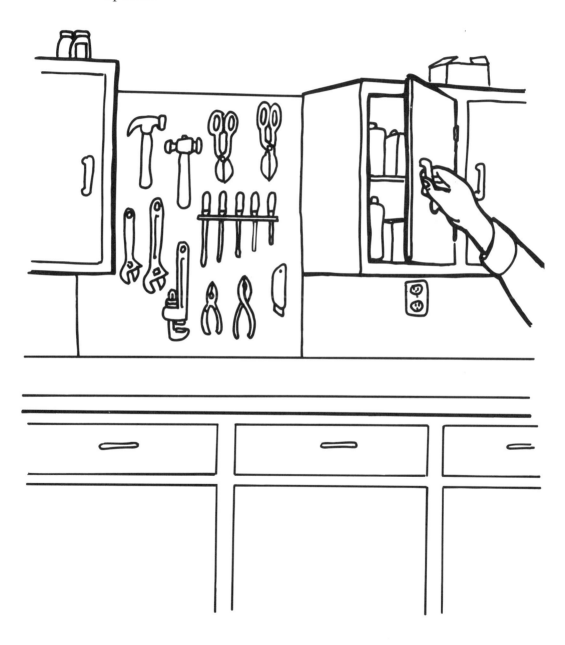

Step 1-3. Be responsible.
Never pour paints, varnishes, or solvents down the drain. Dispose of these materials at a hazardous waste center.

Step 1-4. Be educated.
Always read and follow the manufacturer's instructions on the labels of products you use. Notice any hazard warnings and storage instructions.

Safe Work Habits 5

Step 1-5. Prevent injuries.
Wear safety goggles when using power saws and performing work such as chipping away dried glue.

Step 1-6. Prevent burns.
Wear rubber gloves when working with any caustic chemicals.

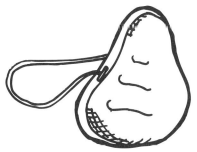

Step 1-7. Avoid inhaling dust and toxic fumes.
Wear a dust respirator when sanding, and work in a well-ventilated area when using toxic chemicals.

Step 1-8. Use electricity wisely.
Use power tools with grounded plugs or two-pronged plugs labeled "double-insulated." Never cut off the ground prong of a plug to use in an ungrounded outlet.

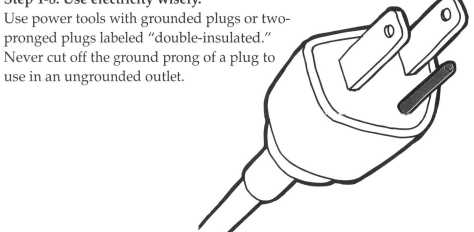

**Step 1-9.
Avoid electric shock.**
Never use any power tool in a damp location.

Step 1-10. Recognize the risks.
Have an ABC fire extinguisher handy and know how to use it.

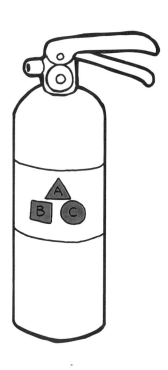

CHAPTER TWO

Tools for the Job

Most furniture repairs can be handled with a basic set of tools, if you are willing to improvise. One example is clamps. You need to apply pressure when gluing. A wide variety of clamps are available, but with a little improvising, you can make do with what you have. Use large rubber bands cut from an old inner tube, twine, masking tape, or wedges to hold parts together until they dry. A rubber mallet is useful to help separate or join parts, but you might get by with a claw hammer wrapped in a few layers of cloth. The following list is a sample of tools and how they can be used for most common repairs to wood furniture.

Screwdriver
Flat blade, small or medium screwdrivers can be used to remove old glue and to pry, as well as to drive screws.

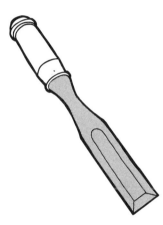

Wood chisel
Small or medium, sharply-honed chisels can be used to cut recesses for hinges and for other wood-shaping jobs.

Masking tape
Masking tape is useful for holding a hinge in place for marking, and it can also be used as a clamp to hold glued parts.

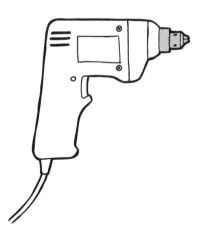

Electric drill
Use an electric drill to drill screw holes and holes for dowels. Use a standard twist bit or a wood bit.

Backsaw
A backsaw is reinforced along the back with a strip of brass or iron. It is often used with a miter box to cut 45- and 90-degree angles. A backsaw has fine teeth for precision cuts.

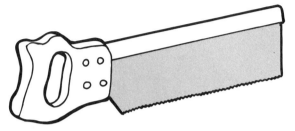

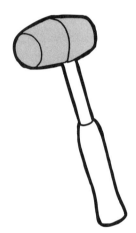

Rubber mallet
A rubber mallet can be used to separate wood joints and tap parts together.

Orbital sander
An orbital sander has a rubber pad for holding sandpaper and is used to sand flat surfaces.

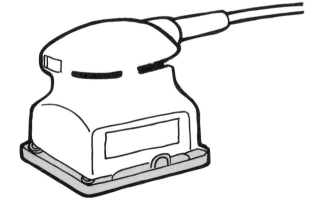

Tools for the Job   11

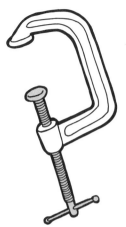

C-clamp
A very useful clamp, the C-clamp is available in a wide variety of sizes. It can double as a vise to hold a part to a work table.

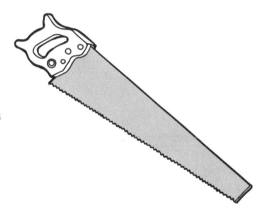

Crosscut saw
A general-purpose saw, the crosscut saw is used for rough cuts across the wood grain.

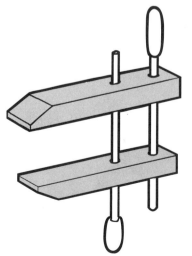

Wood clamp
A wood clamp has threaded rods that adjust the wooden jaws at various angles to hold parts for gluing.

Pipe clamp

A pipe clamp has 1/2- or 3/4-inch pipe cut to length with threads on one end. The adjusting part of clamp is threaded onto the pipe while the holding part of clamp slides along the pipe and locks into place.

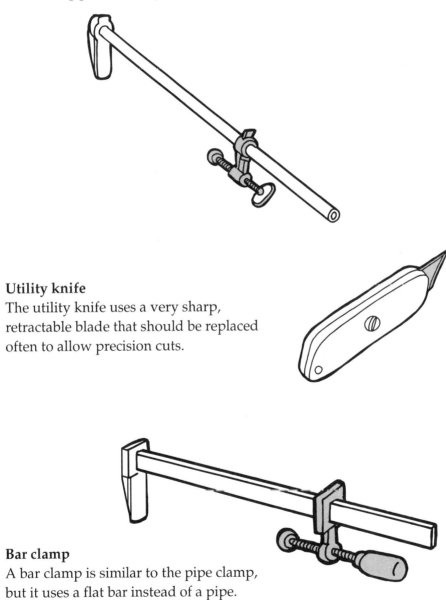

Utility knife

The utility knife uses a very sharp, retractable blade that should be replaced often to allow precision cuts.

Bar clamp

A bar clamp is similar to the pipe clamp, but it uses a flat bar instead of a pipe.

Tools for the Job   **13**

Belt or band clamp
A belt, or band, clamp has a belt that can be several feet long, made of nylon or canvas. This type of clamp is great for holding legs and rungs together for gluing.

White or yellow woodworker's glue
White woodworker's glue works well and dries clear, but yellow glue is slightly stronger, better for sanding, and more resistant to moisture.

Safety goggles
Safety goggles should be worn when you are using power tools or chipping dried glue to protect your eyes from flying debris.

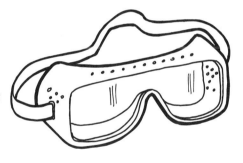

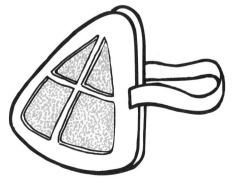

Dust mask
A dust mask with disposable filters provides protection from fine dust when you are sanding.

Respirator
Use a respirator with disposable canister filters when working with materials that produce toxic fumes.

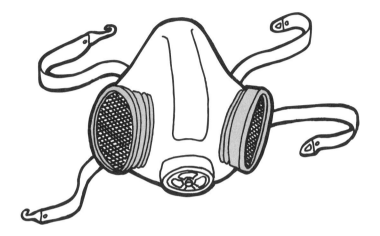

Rubber gloves
Wear rubber gloves when working with caustic chemicals.

Sandpaper
Coarse-grade (60 to 80 grit) sandpaper is used for rough sanding and removing paint. Medium-grade (80 to 120 grit) sandpaper is general-purpose—for removing finishes, leveling shallow dents, and smoothing scratches. Fine-grade (120 to 180 grit) is used for final sanding before applying the finish. Very-fine (220 to 280 grit) is used for the last sanding before applying the finish coat. Extra-fine or super-fine (280 to 600 grit) is used for smoothing fine particles, satinizing varnish and lacquer finishes, and sanding before applying final coat of lacquer.

Steel wool
Medium (grade 1) steel wool is used for removing paint. Fine (grade 1/0, 2/0) steel wool is used for general smoothing and dulling the gloss of varnish or enamel finishes. Extra-fine (grade 3/0) is used with a lubricant for small surface repairs and final smoothing before applying the finish coat. Super-fine (grade 4/0) is used with furniture oil or wax to polish to a high finish.

Your inventory of tools depends on the number of repairs you expect to make and the size or difficulty of these jobs. You can always add to your basic set of tools as the need arises. Just wait until a specific repair requires a special tool, then buy the highest-quality tool that fits your budget. High-quality tools, used and maintained properly, will last a lifetime.

CHAPTER THREE

Work Techniques

There is always more than one way to do a job. How you do it depends on the tools and materials you have available and your expertise. The methods you use should produce the results you desire. Many jobs are probably botched because someone is in a hurry. Lessons learned the hard way prompted sayings such as "Measure twice and cut once." Try to be as precise as you can when measuring and cutting, be neat, and if the glue takes 24 hours to cure completely, set the work aside for 24 hours.

The following pages explain how to perform many of the techniques you might use in repairing wood furniture.

Tools & materials

- ❏ Wood glue
- ❏ Dowels
- ❏ Screwdriver
- ❏ Chisel
- ❏ Sandpaper
- ❏ Pencil
- ❏ Clamps of various types
- ❏ Cheesecloth
- ❏ Toothpicks
- ❏ Pliers
- ❏ Nails
- ❏ Drill and drill bits
- ❏ Masking tape
- ❏ Utility knife
- ❏ Crosscut saw
- ❏ Sawhorses
- ❏ Orbital sander

Work Techniques

Regluing a dowel joint

Glue is preferred over nails or screws for joining parts. Glue does not split the wood and often is much stronger than the wood itself. Dowels are preferred over nails or screws for making a joint. Dowels are wooden pegs that are glued into the sockets of each piece to be joined.

Step 3-1. Removing the old glue.
Carefully scrape the surface of the dowel with the sharp edge of a screwdriver or a chisel to chip away the old glue. Try not to gouge the wood.

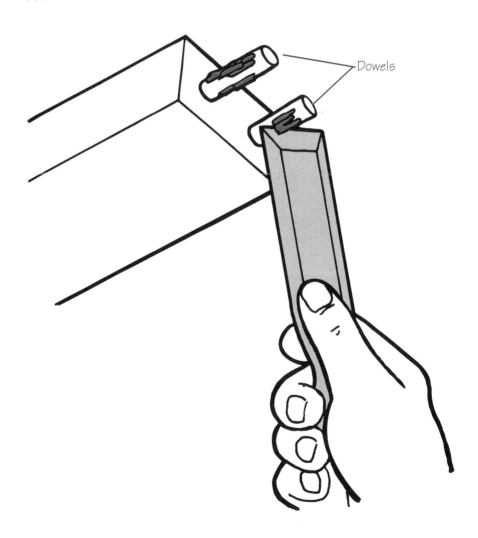

Step 3-2. Sanding the dowel.
Use a strip of medium grade sandpaper to remove any remaining glue. Sand only enough to remove the glue and not enough to change the size of the dowel.

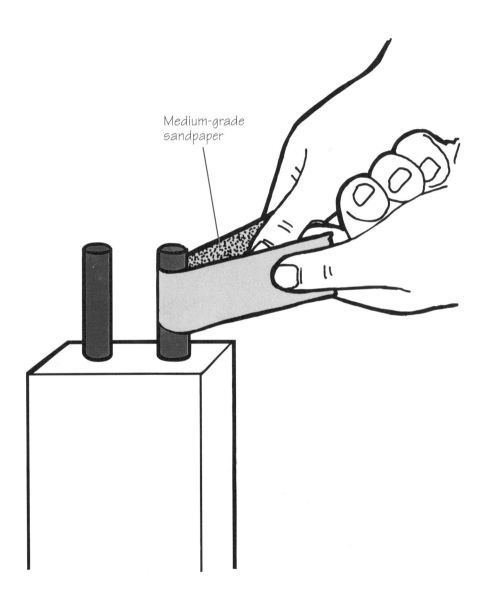

Step 3-3. Sanding the dowel socket.
Use a strip of medium-grade sandpaper wrapped around a pencil or smaller dowel to remove the old glue from the dowel socket. Try to clean as much as possible without changing the shape of the socket.

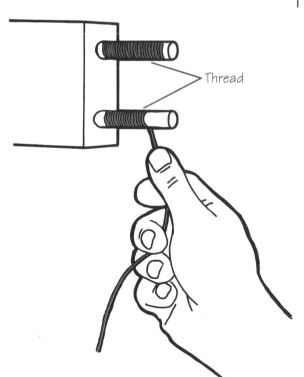
Thread

Step 3-4. Making a dry run.
Before applying any glue, fit the pieces together to check that the joint is snug. Make sure you have suitable clamps available once the glue has been applied. If the joint is a little loose, mix a bit of sawdust with the glue. If the fit is sloppy, spread a thin layer of glue on the dowel and wrap it with thread or a strip of cheesecloth or cotton cloth. Wait for the glue to dry, then make another dry run.

20 Fixing Furniture

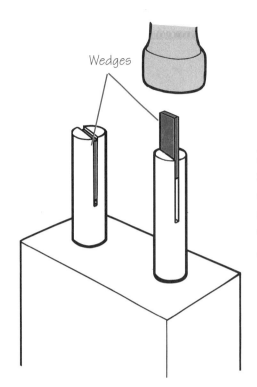

Step 3-5. Using a wedge.
If the dowel is very loose, saw a slot a little more than halfway down the center of the dowel, tap a wedge into the slot to expand the dowel, then try the joint again for a good fit. If the joint is good, separate it, apply the glue, assemble the joint, and install the clamp.

**Step 3-6.
Applying the glue to the socket.**
After you have made a dry run, drip a little glue into the dowel socket. Then use a toothpick to spread a thin layer of glue around the walls of the socket. Don't fill the socket with glue.

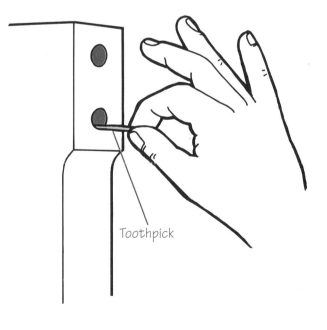

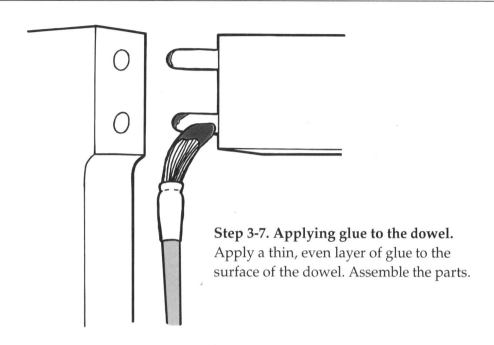

Step 3-7. Applying glue to the dowel.
Apply a thin, even layer of glue to the surface of the dowel. Assemble the parts.

Step 3-8. Apply the clamp.
Make sure the parts are properly aligned. Protect the wood finish where the clamp will be placed. Apply the clamp and tighten until you see a little glue ooze from the joint. Don't make the clamp too tight or you will squeeze all the glue from the joint. Remove excess glue with a damp cloth. Let the work set for the time recommended by the glue manufacturer.

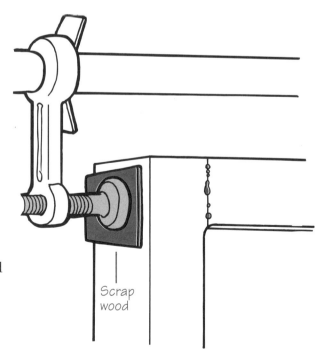

Repairing broken dowel joints

When a joint becomes loose, you can try adding new glue and clamping until dry, but for a better bond you'll probably have to take the pieces apart and remove all of the old glue.

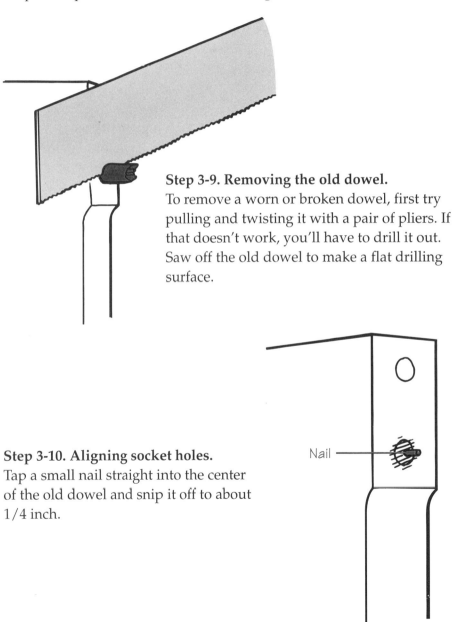

Step 3-9. Removing the old dowel.
To remove a worn or broken dowel, first try pulling and twisting it with a pair of pliers. If that doesn't work, you'll have to drill it out. Saw off the old dowel to make a flat drilling surface.

Step 3-10. Aligning socket holes.
Tap a small nail straight into the center of the old dowel and snip it off to about 1/4 inch.

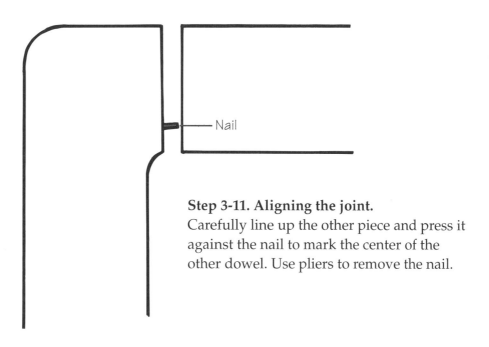

Step 3-11. Aligning the joint.
Carefully line up the other piece and press it against the nail to mark the center of the other dowel. Use pliers to remove the nail.

Step 3-12. Drilling the sockets.
Estimate the depth of the socket and add about 1/4 inch more to drill into solid wood. Choose a drill bit that is the same size as the replacement dowel. Measure up from the end of the bit and wrap masking tape around it to mark where you should stop drilling. Drill as straight as possible into the old dowel, stopping at the depth marked by the tape. Drill out both sockets and blow out any dust.

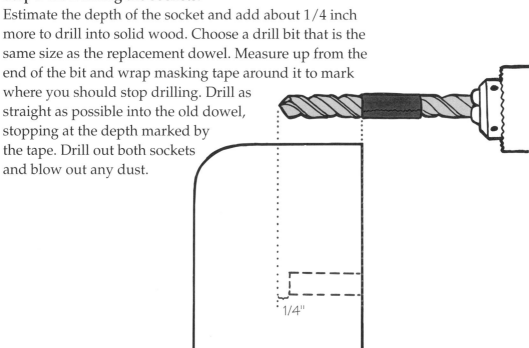

Step 3-13. Cutting the dowel.

Measure and cut the new dowel to fit the combined depth of both sockets. Use coarse sandpaper to sand a slight bevel on both ends of the dowel. Make small grooves around the dowel by gently crimping it with the notched jaws of a pair of pliers. The beveled ends and the grooves prevent excess glue from being trapped when the dowel is pressed into the socket and provide a larger surface area for gluing.

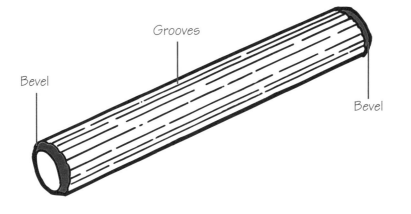

Step 3-14. Making a dry run.

Assemble the joint and make sure you have your clamps handy. Make any necessary modifications now for a proper fit.

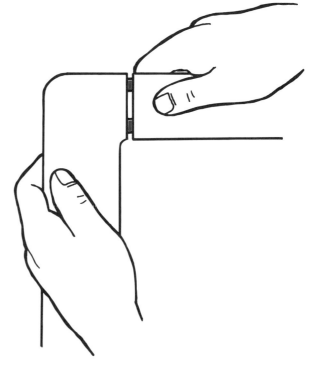

Step 3-15. Gluing the joint.
Apply white or yellow glue to the dowel sockets and dowel. Reassemble the joint and install the clamps. Wipe off any excess glue and set the parts aside overnight to dry.

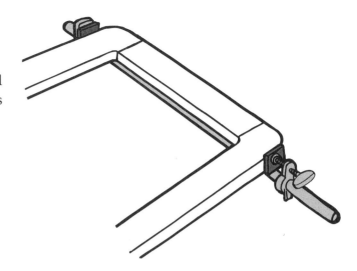

Repairing hinges

Step 3-16. Repairing split hinge beds.
When a split develops between the screw holes in a hinge bed, separate the split enough to brush or blow out any loose debris. Then use a toothpick to force glue into and along the length of the split. Install clamps to pull the split back together, wipe off the excess glue, and let it dry overnight.

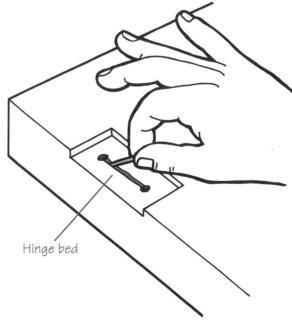

Step 3-17. Repairing loose hinge screws.
Remove the screw. Dab the ends of two or three toothpicks in wood glue, and gently tap them, one at a time, into the hole. Use a utility knife to cut off the ends flush with the surface. Be careful if you must pull the knife blade toward yourself. Let the glue dry, then reinstall the screw. For a large hole, first drill out the hole with a bit the size of a dowel. Dip the end of the dowel in glue and tap it into the hole. Let it dry, then cut off the excess with a sharp chisel or saw. Now use a small bit to drill a pilot hole for the screw and reinstall the screw.

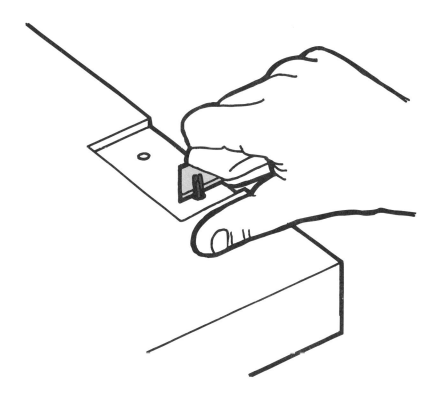

Using a crosscut saw

Not as common as it once was because of the popularity of the circular saw, the crosscut saw is designed for quick, rough cuts across the grain of the wood. Be careful—you can scrape your thumb or knuckles when you do not start the cut properly or place the wood on a stable platform.

Step 3-18. Preparing to saw.
After you have marked the cut with a carpenter's pencil, place the board on a solid, low work platform—for example, two sawhorses or a work bench. Place the cut mark close to the sawhorse, leaving enough room for the saw to clear. Place one knee and one hand on the board. Fold your thumb under your index finger to protect it from accidental slips. Now lean over so that your upper body weight holds the board firmly against the sawhorse. You should be able to sight down the edge of the board where the cut begins.

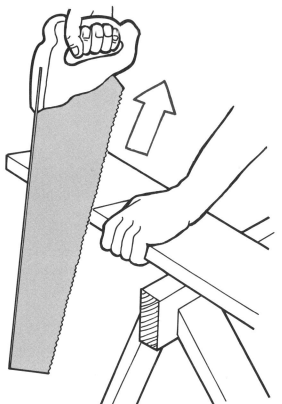

Step 3-19. Starting the cut.
Grip the saw handle firmly with your other hand. Hold the blade almost perpendicular to the board, placing the cutting edge against the cut mark on the board. Your arm and shoulder should be directly over the cut mark. Next, place the teeth of the blade against the wood, holding slight pressure. Slowly pull the blade up toward you a few times until you have cut a deep, solid notch in the wood.

Step 3-20. Sawing the board.
After you cut a starting notch, hold the saw so that the blade is at about a 45-degree angle to the wood. Using a little pressure on the downstroke, saw through the board. Cut on the downstroke, letting the teeth do the work.

Step 3-21. Completing the cut.
When you have cut to within about an inch of the edge, steady the waste end of the board with your hand. Hold the waste end as straight as possible, otherwise it will pinch the blade. Hold the blade vertically and use short up and down strokes to finish the cut.

Using an orbital sander

The orbital sander is a finishing tool. It is not designed to remove material. Some models have a lever to switch from orbital motion to straight-line sanding. Sandpaper is fastened to the pads by clamps, sliding clips, or rods. The most common type of sander uses lever-operated clamps.

Step 3-22. Attaching the sandpaper.
Push in and release the levers on the sander to open the clamps. Use precut sander sheets or cut a piece of sandpaper as wide as, but slightly longer than, the pad. Insert one end of the sandpaper under one clamp; close and lock the clamp. Pull the sandpaper tight across the pad and insert the end under the remaining clamp. Close and lock the clamp.

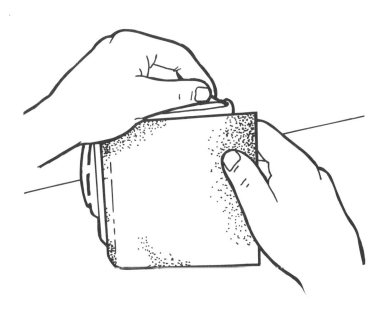

Step 3-23.
Using the sander.
Hold the sander firmly and turn it on. Move the sander over the surface without pressing down. Let the weight of the sander do the work. Work in smooth, even passes, lifting the sander after each pass. Check the condition of the sandpaper periodically and replace it when it becomes worn.

Using a chisel

Your chisel probably has a plastic handle designed to withstand medium taps with a hammer or mallet. The chisel blade should be sharp.

Step 3-24. Starting the cut.
To cut out a recessed area, hold the chisel straight up and down, with the beveled side in. Cut along the outer edges of the area.

32 Fixing Furniture

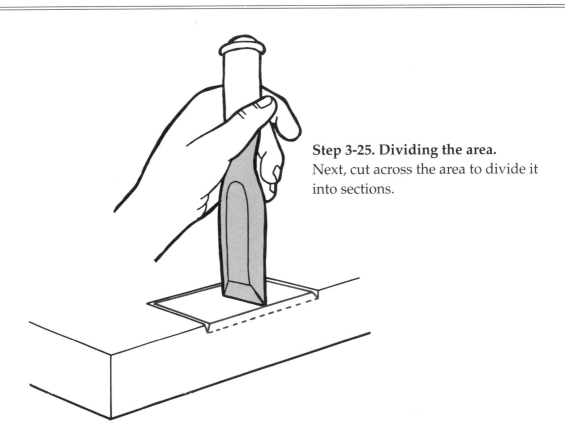

Step 3-25. Dividing the area.
Next, cut across the area to divide it into sections.

Step 3-26. Removing the waste.
When the area is closed, hold the chisel with the beveled side down and remove the sections. If the area to be cut is open on one end, peel the waste off with the bevel side up.

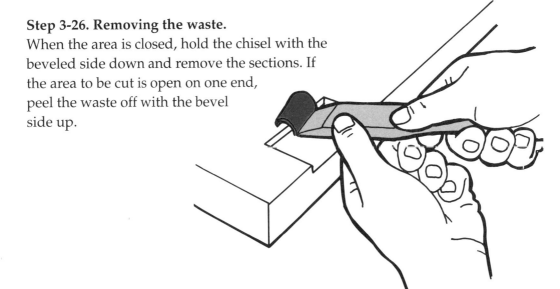

Using clamps

Several types of clamps are available for repairing furniture, including C-clamps, hand screw clamps, web clamps, and pipe or bar clamps. However, you can complete most repairs by using common household items, such as heavy twine or masking tape.

Step 3-27. Using masking tape.
When using masking tape as a clamp, apply the glue, press the parts together and wipe off the excess glue. Now wrap the tape tightly around the parts to hold them in place. Set the work aside overnight to dry.

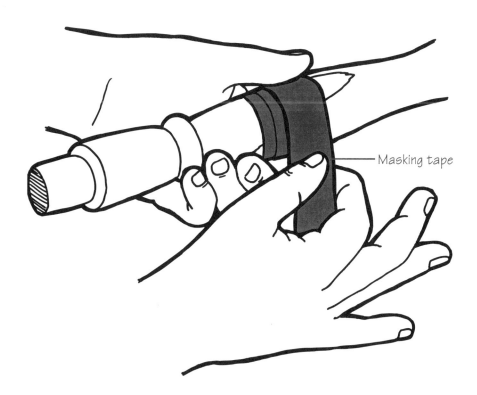

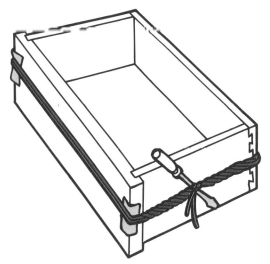

Step 3-28. Using twine.
Try to use twine or cord that doesn't stretch. A heavy cotton clothesline works well. Pad the furniture with folded cloth or cardboard to prevent the twine from marring the finish. Wrap the twine twice around the parts and tie a knot. Slip a screwdriver, dowel, or stick between the strands of twine and twist it like a tourniquet. Continue twisting until you have the desired pressure. Now brace the screwdriver against the furniture and allow the glue to dry.

Step 3-29. Using C-clamps.
C-clamps typically have openings from 1 to 8 inches wide. The foot of the clamp has a ball joint that is designed to swivel in order to fit work that is slightly angled. Always use a strip of scrap wood as a pad to protect the work's finish and to distribute the pressure of the clamp evenly.

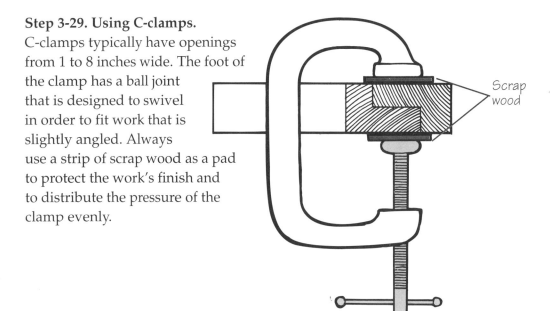

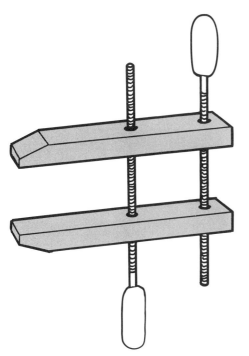

Step 3-30. Using hand screw clamps.
The hand screw clamp has wood jaws that normally do not need pads to protect the furniture. The screws are threaded through pivots in the jaws so that the jaws can be adjusted to different positions.

Step 3-31. Using web clamps.
The web clamp uses a canvas or nylon strap up to 15 feet long, with a locking device to hold the work. This clamp can be used to hold irregular shapes or to pull several joints together at the same time. To use the web clamp, thread the end of the strap through the lock, loop the strap around the furniture, then pull end of the strap to draw the loop in place. Place cloth or cardboard pads between the strap and the furniture to protect the finish, then crank or ratchet the lock to tighten the strap.

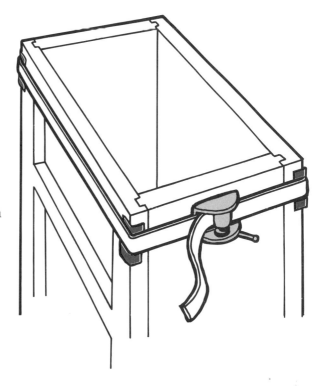

Step 3-32. Using bar clamps.

Bar clamps have jaws mounted on a flat steel bar up to 4 feet long. One end has a movable jaw and the other end has a fixed jaw operated by a cranking screw. To use this clamp, position the fixed jaw against the side of the work, slide the movable jaw against the other side, then turn the crank to apply pressure to the work. Use pads to protect the surface of the work.

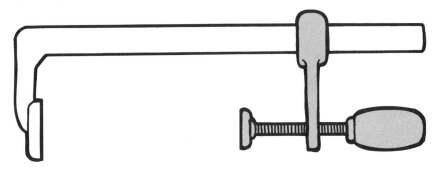

Step 3-33. Using pipe clamps.

Pipe clamps are similar to bar clamps and are useful for large furniture repairs like split or cracked table tops. You buy the jaws to fit either 1/2- or 3/4-inch threaded steel pipe. One jaw is movable; the other jaw is fixed and is screwed onto one end of the pipe. This clamp is used like a bar clamp except that the pipe can be cut, or, by using couplings, extended to any length you need.

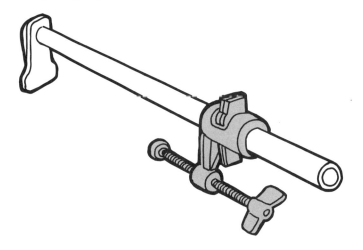

Step 3-34.
Using improvised C-clamps, homemade clamps, string, wedges, pipe wrenches, and thread spools.
When C-clamps are too small, try hooking the jaws together, then tightening the screws, or try connecting the jaws with a loop of heavy twine or wire cable. For smaller jobs, run a long bolt through two spools or a couple of washers and tighten with a wing nut. For larger jobs, nail a couple of braces on each side and tap in a wedge.

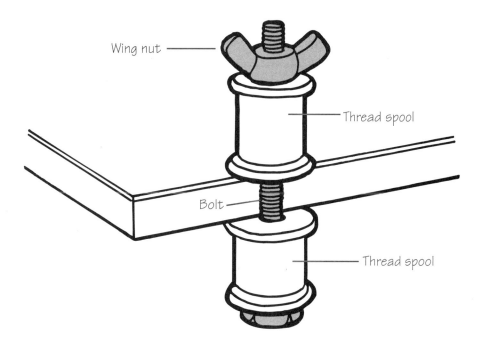

CHAPTER FOUR

Repairing Chairs

A chair is normally thought of as just something to sit on, but, much like a bridge, it is quite a piece of structural engineering. Chairs usually don't weigh much, but they often support a couple hundred pounds that might be rocking or tilted back on two legs. This strength is possible because of the bracing between the legs, back, and seat.

You can easily understand that one loose joint puts a much greater strain on additional joints, causing them to weaken and separate. Major problems can be avoided by inspecting and repairing the minor ones as soon as they are noticed. Most minor repairs can be completed with little more than some glue, heavy twine, and a screwdriver.

You can separate the joints with a lever arrangement made from a padded screwdriver and a piece of scrap wood, a cloth-covered hammer, or a rubber mallet. If you take the chair completely apart, set the pieces aside in a logical order so you know which part goes where when you put it back together.

Repairing Chairs   39

A typical wooden chair is constructed in one of two styles: platform or frame. The platform chair has a seat supported by legs braced by foot rails, with a back that fits into sockets in the top of the seat. The frame chair has a back that includes the two back legs, joined to a frame consisting of the front two legs and the seat.

The most common methods of joining the pieces of a wooden chair together are dowel joints or mortise-and-tenon joints. With a dowel joint, the pieces are joined with a dowel fitted into a socket. The mortise-and-tenon joint is similar—the dowel part, called a tenon, is rectangular and fits into an opening called a mortise.

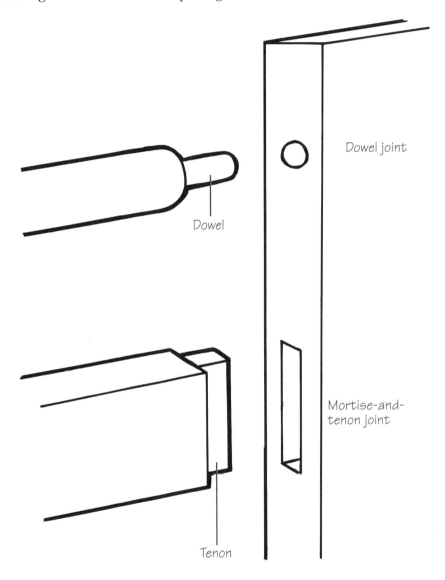

Tools & materials

- ❏ Screwdriver
- ❏ Wood glue
- ❏ Clamps
- ❏ Thread or cheesecloth
- ❏ Utility knife
- ❏ Masking tape
- ❏ Coarse sandpaper
- ❏ Sanding block
- ❏ Padded hammer or rubber mallet
- ❏ Iron
- ❏ Hose clamps

Platform chairs

Steps 4-1 through 4-4 show you how to reglue loose foot rails.

Step 4-1. Separating the joint.
Try to pull the joint apart by hand gently. If this method fails, look for hidden fasteners such as brads or small nails covered with wood filler, and remove them. If the joint is only slightly loose, separate it as far as you can without damaging another joint, use the tip of a screwdriver to scrape off the old glue, apply fresh glue, and clamp the joint using a tourniquet of twine or a web clamp.

Step 4-2. Separating a stubborn joint.
To separate a stubborn joint, apply steady pressure with a lever made from a padded screwdriver or hammer handle and a piece of scrap wood.

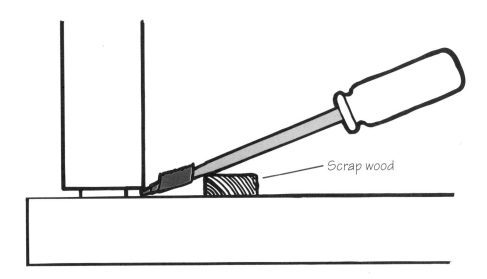

Step 4-3. Regluing a very loose joint.
If the joint is very loose, wrap the dowel with thread, apply glue, and let dry. Then try it for size. If it is still too loose, fold a couple narrow strips of cheesecloth over the end of the dowel, apply the glue, and assemble the joint.

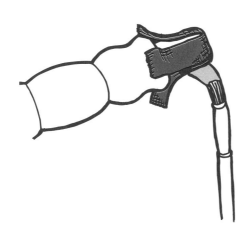
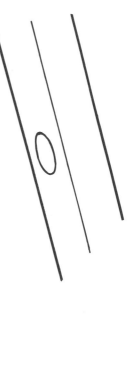

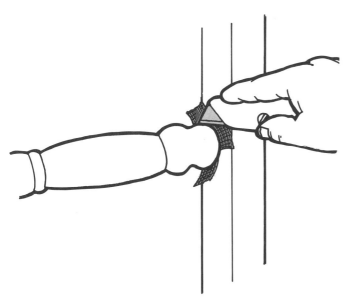

Step 4-4. Finishing the job.
Install the clamp, trim off the excess cloth with a utility knife, and wipe off the excess glue.

Steps 4-5 through 4-7 explain how to reglue the legs.

Step 4-5. Separating the leg assembly.
Before separating the leg assembly from the seat, wrap a piece of masking tape around one of the legs and mark it; right front (RF), left front (LF), etc., to avoid any confusion when you reassemble. Try to separate the legs from the seat by hand, being careful not to disturb the joints in the foot rails. If this method fails, use a padded screwdriver and a piece of wood for leverage, or place the legs in a padded vise one at a time, and tap the bottom of the seat near each leg.

Scrap materials

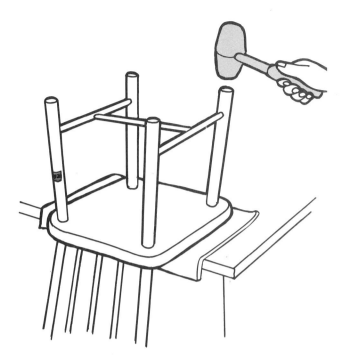

Step 4-6. Gluing the legs.
Use the tip of a screwdriver or coarse sandpaper to remove the old glue from the joints. Fit the legs back into the seat according to your labels, and check for a good fit. Use thread or cloth and glue to tighten up any loose joints and allow to dry. After you are satisfied with the joints, separate them, and, with the seat upside down on a firm surface, apply the glue and tap the legs into their sockets with a hammer or rubber mallet.

**Step 4-7.
Applying weight to set the glue.**
Set the chair on a smooth, flat surface, and place a stack of books, a bucket of water, or anything heavy on the seat. Allow the glue to dry overnight.

Steps 4-8 through 4-12 show you how to reglue the back of a chair.

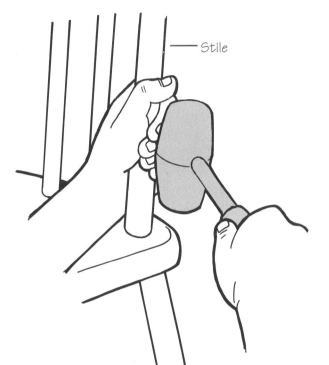

Step 4-8.
Removing the back of a chair.
To remove the back of the chair, hold the first stile firmly with one hand and tap near the joint in the seat. Use a cloth-covered hammer or a rubber mallet. Continue with the second stile and the third until you have the back free.

Step 4-9. Removing the old glue.
Use the tip of a screwdriver and coarse sandpaper to remove the old glue from the dowels and sockets. Fit the back of the chair into the sockets and check for any loose joints. Repair any loose joints with thread or cloth and glue. Let the repairs dry, then check again for snug joints.

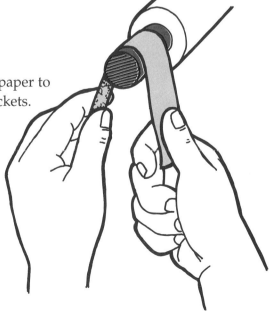

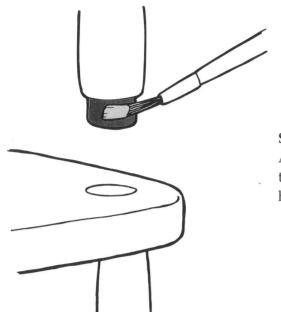

Step 4-10. Gluing the back.
Apply glue to the sockets and dowels and tap the back in place with a padded hammer or rubber mallet.

Step 4-11. Installing a bar clamp.
If you have a bar or pipe clamp, clamp the chair from the bottom of the seat to the top of the back and tighten the screw to seat the dowels in their sockets. Wipe off the excess glue with a damp cloth. Let dry overnight.

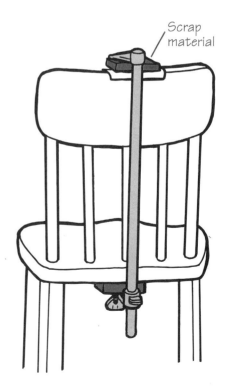

Step 4-12. Using a web clamp or twine.
If a bar clamp is not handy, run one or two web clamps under the seat and around the back. Position the ratchets at the back of the chair and tighten. Use heavy twine in a tourniquet fashion if other clamps are not available. Wipe off any excess glue with a damp cloth and let dry overnight.

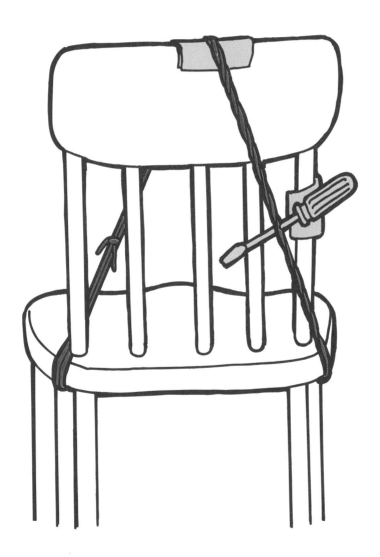

Repairing Chairs 49

Frame chairs

Steps 4-13 through 4-16 cover repairing loose legs.

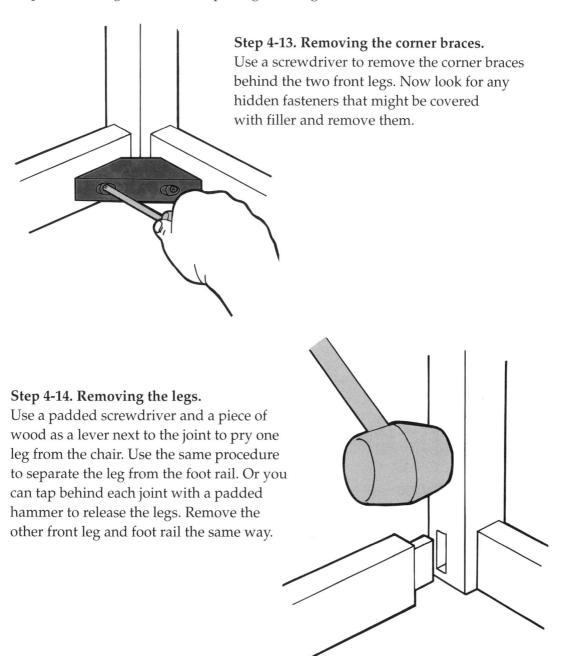

Step 4-13. Removing the corner braces.
Use a screwdriver to remove the corner braces behind the two front legs. Now look for any hidden fasteners that might be covered with filler and remove them.

Step 4-14. Removing the legs.
Use a padded screwdriver and a piece of wood as a lever next to the joint to pry one leg from the chair. Use the same procedure to separate the leg from the foot rail. Or you can tap behind each joint with a padded hammer to release the legs. Remove the other front leg and foot rail the same way.

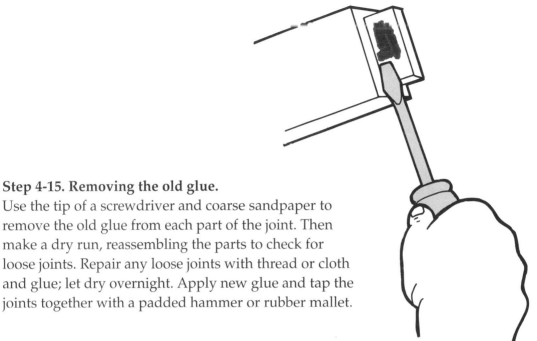

Step 4-15. Removing the old glue.
Use the tip of a screwdriver and coarse sandpaper to remove the old glue from each part of the joint. Then make a dry run, reassembling the parts to check for loose joints. Repair any loose joints with thread or cloth and glue; let dry overnight. Apply new glue and tap the joints together with a padded hammer or rubber mallet.

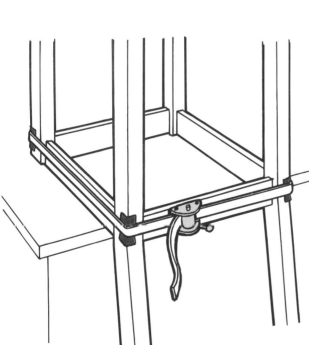

Step 4-16. Installing the clamp.
Use a web clamp or make a tourniquet clamp from heavy twine to pull the joints together. Wipe off the excess glue with a damp cloth and allow to dry overnight.

Repairing Chairs 51

Steps 4-17 through 4-21 show you how to repair a loose back on a frame chair.

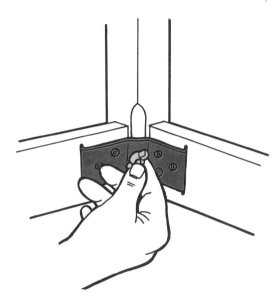

**Step 4-17.
Removing the corner braces.**
Use a screwdriver to remove the corner braces behind each back leg. Look for any hidden fasteners and remove them.

Step 4-18. Removing the back from the seat.
With a padded hammer, tap the back of the chair just above the seat to free the joint. Now go to the foot rail and use a padded screwdriver and a piece of wood to pry the back from the foot rail. You also could try tapping the back from the foot rail with a padded hammer. Repeat the steps to separate the joints on the other side.

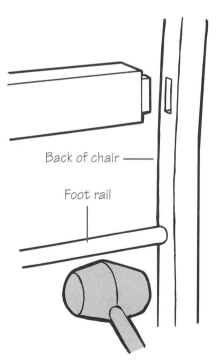

Back of chair

Foot rail

Step 4-19. Removing the old glue.
Use the tip of a screwdriver and coarse sandpaper to remove as much of the old glue as possible. Reassemble the chair to check for loose joints and repair any with thread or cloth and glue. Let repairs dry overnight.

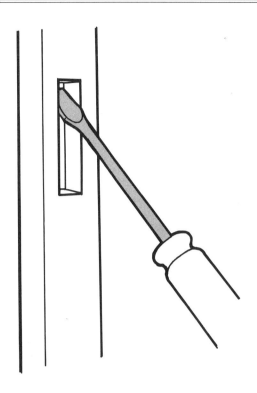

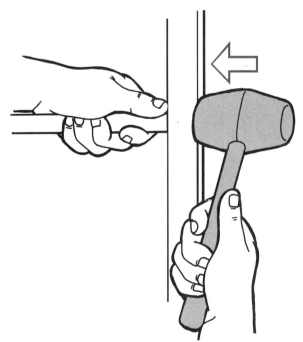

Step 4-20. Applying the glue.
Apply new glue to each part and reassemble the chair. Tap the joints together with a padded hammer. Place the chair upright on a smooth, flat surface to be sure all of the parts are aligned properly.

Step 4-21. Installing the clamp.
Use a couple of web clamps or heavy twine in a tourniquet to pull the joints together. Wipe off the excess glue with a damp cloth and set the work aside to dry overnight.

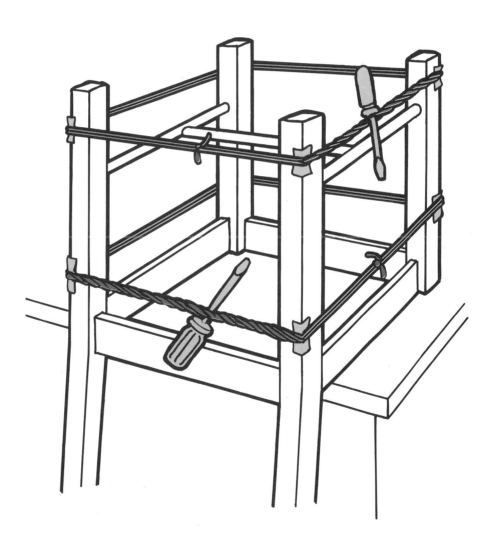

54   Fixing Furniture

Wobbly chairs are caused by uneven legs as a result of wear, a missing glide, or faulty repairs. The repair usually can be done in a matter of minutes. Steps 4-22 through 4-24 show you how.

Step 4-22. Locating the faulty leg.
First check the bottom of the legs for a missing glide. If the legs have no glides, or none is missing, place the chair on a smooth, flat surface and rock it back and forth. You should find that one or possibly two legs are shorter and are causing the wobble. The longer legs must be shortened to this length. Mark the long legs with masking tape.

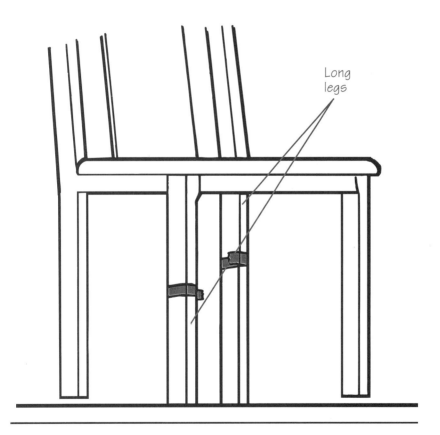

Repairing Chairs 55

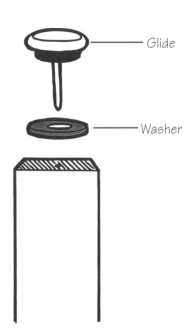

Step 4-23. Installing new glides.
If one of the glides is missing, use the tip of a screwdriver or a kitchen knife to remove the old ones. Tap in a new set with a hammer. If all of the glides are still there, remove the one on the shorter leg, place a metal washer on the bottom of the leg and tap the glide back in place.

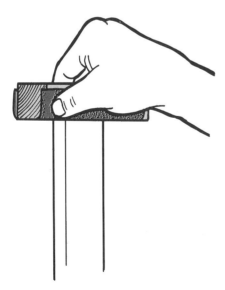

Step 4-24. Sanding.
If the legs have no glides, place the chair upside down on a level surface and use a sanding block with coarse sandpaper to sand down the longer legs to match the shorter legs. Sand off a little at a time and recheck for the wobble often. Otherwise you could add to the problem by sanding off too much. Be sure to sand at the correct angle so that the leg stands flat on the floor.

Bentwood furniture

When a split develops in bentwood furniture, you can make the repair with a couple of automotive hose clamps. Steps 4-25 through 4-27 show you the procedure.

Step 4-25. Softening the wood.
Use the tip of a knife to remove any residue from inside the split. Wet a cloth with water and wrap it around the split. Hold a hot iron against the cloth to steam the split. Rewet the cloth as needed and move the iron around to completely steam the split until the break becomes pliable.

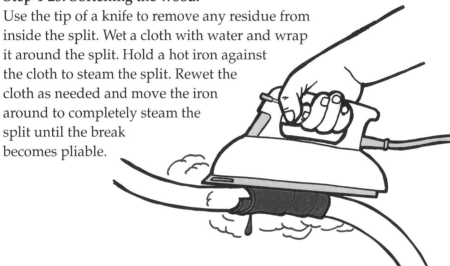

Step 4-26. Shaping the wood.
Press the split closed and wrap it with a few turns of masking tape. Now use a couple of hose clamps to hold the split in place. Tighten the clamps only enough to close the break, not enough to mark the wood. Let the wood dry a day or two.

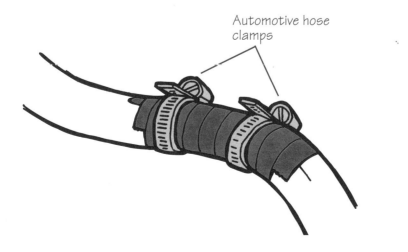

Automotive hose clamps

Step 4-27. Applying the glue. Carefully lift the split and use a toothpick to spread the glue evenly inside the break. Press the split back together by hand and wipe off the excess glue with a damp cloth. Place a little waxed paper over the split and install the hose clamps as before. Let the repair dry overnight.

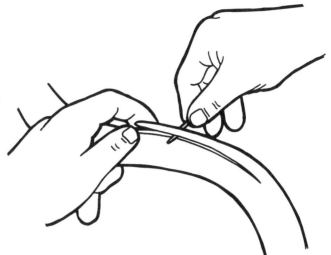

CHAPTER FIVE

Repairing Tables

Heavy tables usually have the table tops fastened to a wood frame, called an apron, that is supported by the legs. The legs might be permanently fastened to the apron by a wood joint or fastened by a bolt on the leg fitted through a corner brace in the apron and secured by a wing nut.

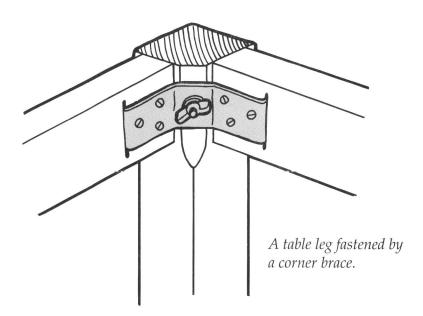

A table leg fastened by a corner brace.

Lighter-duty tables might have the legs fastened directly to the table tops. Table tops can be one solid piece, or they can be made up in sections, like extension and drop-leaf tables. Wood joints that become loose can be reglued using the same procedure you use with chairs. However, in tables, dowel joints are not as common as mortise-and-tenon joints or dovetail joints. Use care when separating the joints if you're not sure which type they are.

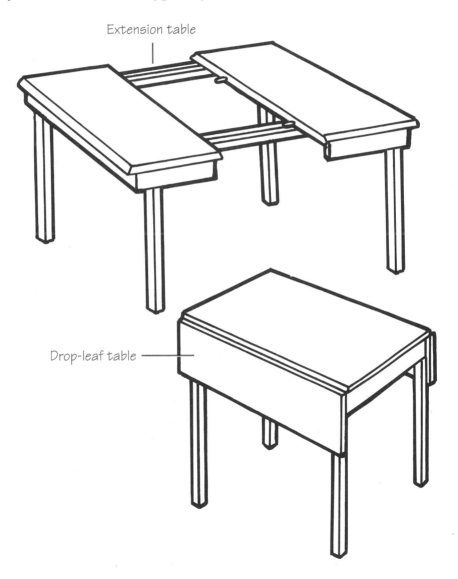

Hinges on drop-leaf tables can be repaired, and balky mechanical parts on extensions can usually be cured by simple cleaning and lubricating.

Tools & materials

- Screwdriver
- Padded hammer or rubber mallet
- Sandpaper
- Old paintbrush
- Wood glue
- Cheesecloth or thread
- Clamps and heavy books
- Wrench
- Padded vise
- Adjustable pliers
- Dowel
- Drill and drill bits
- Masking tape
- Clean cloth
- Old toothbrush
- Mineral spirits or turpentine
- Silicone spray, paraffin, or candle wax
- Toothpicks
- Utility knife
- Iron
- Wax paper

Repairing a loose wood joint

Step 5-1. Removing the table top.

Turn the table upside down on an old blanket or piece of scrap carpet. Locate the screws holding the top to the apron. The screws might be in metal clips or angled holes in the side of the apron. Remove the screws and lift off the legs and apron.

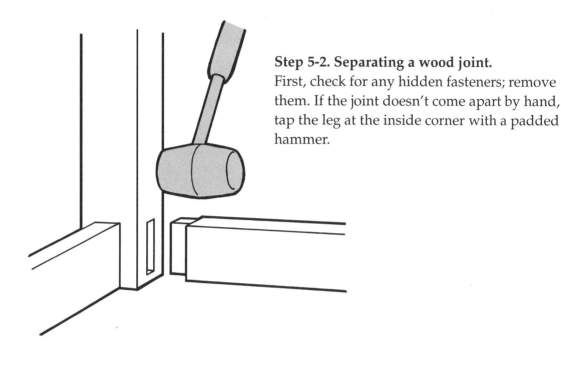

Step 5-2. Separating a wood joint.
First, check for any hidden fasteners; remove them. If the joint doesn't come apart by hand, tap the leg at the inside corner with a padded hammer.

Step 5-3. Removing the old glue.
Use the tip of a screwdriver and coarse sandpaper to scrape away the old glue. Use an old paintbrush to brush any debris from the mortise-and-tenon joint. Make a dry run, putting the joints back together to check for proper fit. Repair any loose joints with thread and glue, cloth and glue, or wedges.

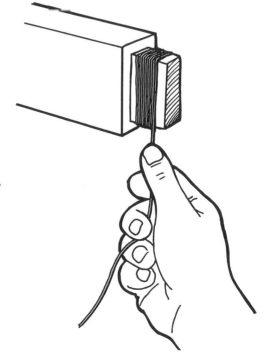

Step 5-4. Installing the clamp.

Apply fresh glue and reassemble the joint. Tap the joint together with a padded hammer. Now set the framework right side up on a smooth, flat surface and check for any wobble. Use one or two pipe clamps to pull the joint together—not too tight, just enough to squeeze out a little glue. If you don't have a pipe clamp, try a web clamp or heavy twine in a tourniquet to close the joint. Set the work aside overnight to dry, then install the table top.

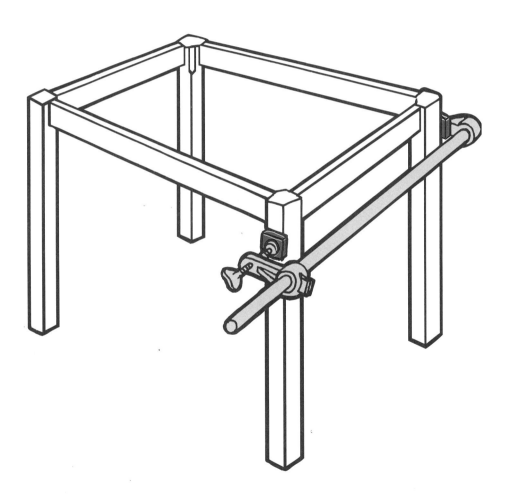

Step 5-5. Repairing a bolted joint.

Sometimes the wing nut or regular nut has worked loose. Use your hand to tighten a wing nut, a wrench to tighten a regular nut. A metal washer should be between the nut and the corner brace. The other end of the bolt has larger threads and is screwed into the wood leg. If the nut fails to tighten the joint, you need to repair the leg. Turn the table upside down on an old blanket and remove the leg.

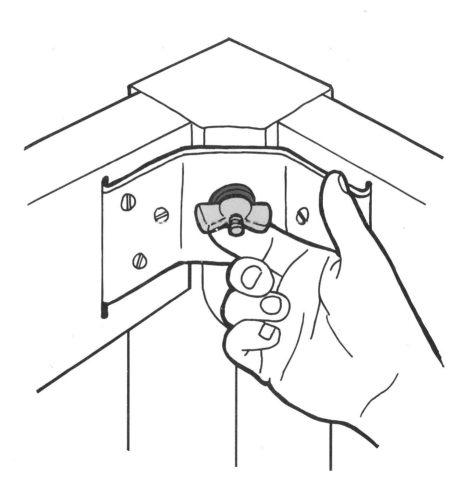

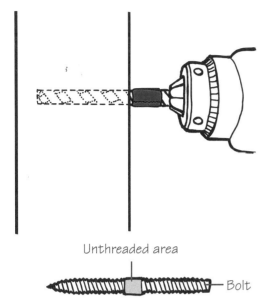

Unthreaded area — Bolt

Step 5-6. Repairing the hole.
Use a padded vise or wedge the leg on a firm surface so you have access to the bolt. The middle of the bolt should not have any threads. Grip this part with pliers and turn the bolt counterclockwise to remove it. Now you have to plug the hole. Choose a dowel slightly larger than and about an inch longer than the hole in the leg. Use a drill bit the same size as the dowel. Determine the depth of the hole and mark this depth on the bit with masking tape. Next, drill out the hole following the same angle and depth of the old one. Blow or shake out any debris left in the hole.

Step 5-7. Installing the dowel.
To keep from trapping the glue in the hole and to increase the gluing surface area, bevel the end of the dowel slightly with sandpaper, then crimp shallow grooves in it with pliers. Apply the glue and tap the grooved end of the dowel into the hole with a hammer. Saw the excess part of the dowel flush with the leg and let the glue dry.

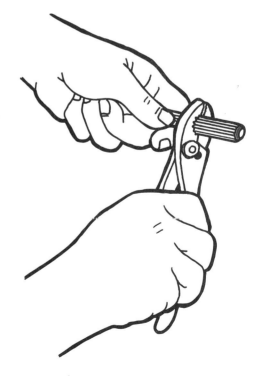

Step 5-8. Installing the bolt.
Use a hammer and nail to tap a starting point in the center of the dowel. Use a tape measure to measure the width of the bolt and the end of the bolt with the larger threads. Choose a drill bit about 1/8-inch smaller than the bolt and mark the bit with masking tape to a depth about 1/4 inch less than the length of larger threads. Now start the bit at the center of the dowel and drill at the same angle to the depth of the masking tape on the bit. Grip the middle of the bolt (no threads) with locking pliers and screw the bolt clockwise into the hole. Stop when the unthreaded part reaches the leg. Reinstall the leg on the apron and fasten it with a washer and wing nut or regular nut.

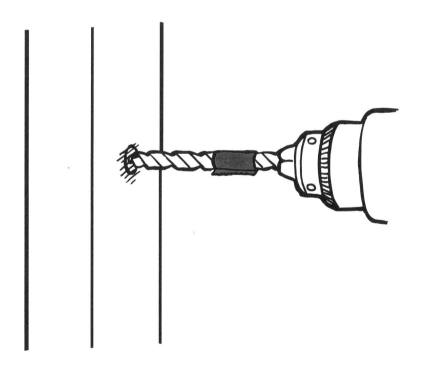

Repairing a balky extension table

Step 5-9. Cleaning wood runners.

Extension tables become difficult to open when the runners are clogged with dirt and old lubricant. Spread the top apart as far as you can to allow access to the runners. Use the tip of a small screwdriver to dig out dirt and caked lubricant from the runners. Now use a cloth soaked with mineral spirits, or turpentine and an old toothbrush, to scrub away any remaining dirt or grease. Wipe the runners dry with a clean cloth.

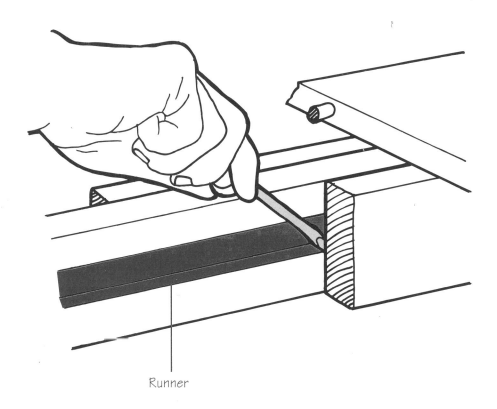

Runner

Step 5-10. Lubricating runners.

Lubricate the runners with a silicone spray or use a piece of paraffin or a wax candle to apply a thin layer of wax to the sliding surfaces of the runners. Slide the top together a few times to spread the wax along the runners.

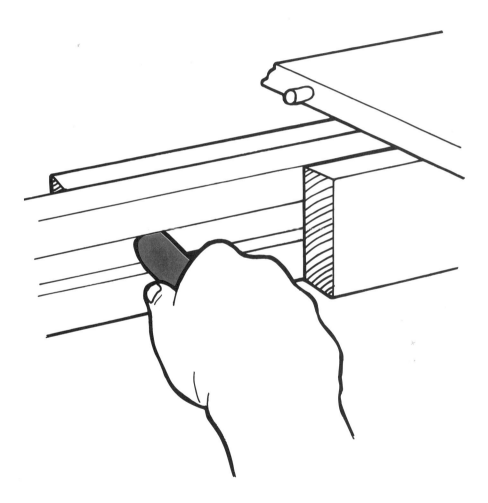

Step 5-11. Repairing a balky sprocket.
Open the table far enough to allow access to the sprocket. Check the sprocket to see if it is loose. If it is, try to tighten the screw with a screwdriver. If the screw won't tighten, the screw hole has probably been stripped. Remove the screw and sprocket and tap a couple of toothpicks coated with glue into the hole. Cut off the excess ends of the toothpicks with a utility knife and reinstall the sprocket and screw. If the screw was tight but the sprocket is dirty, remove the screw and sprocket. Scrub the sprocket with an old toothbrush and mineral spirits to remove the old grease. Reinstall the sprocket. Use the toothbrush to clean old grease from the tracks, then spray the sprocket and tracks with a silicone lubricant or sprinkle them with powdered graphite.

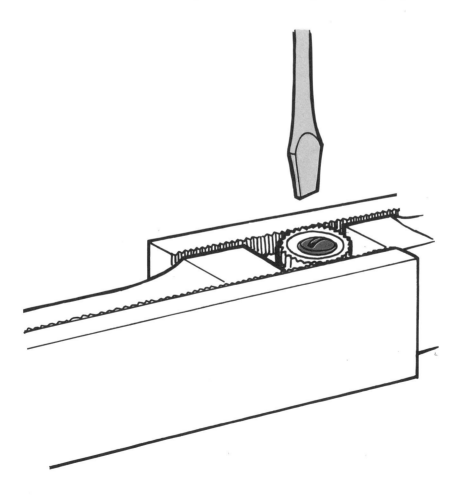

Regluing veneer

Because of the high costs of solid wood, most tabletops, desks, and other large surfaces have a thin layer of decorative wood, called a veneer, glued to the top of a less-expensive wood. The most common problem found with a veneer is caused when the old glue fails and a blister forms or the veneer lifts near the edge of the piece. Because the old glue softens with heat and moisture, you might be able to make the repair with only an electric iron and a towel.

Step 5-12. Repairing a small blister.
Place a dish towel over the blister to protect the finish, and press down on the blister with an iron set on moderate heat. After a few minutes, the blister should flatten. Once the blister has flattened, weight it down with several heavy books overnight.

Step 5-13. Repairing a large blister.

If the blister is large, but not broken, try the method in Step 5-12 first. It might not work because the veneer has probably swelled. In this case, use a utility knife with a fresh blade to make a slit along the grain the full length of the blister. Place wax paper over the slit and cover with a dish towel. Press down on the blister with an iron set on moderate heat. Check the blister frequently. You should see the blister flatten, with areas along the edges of the slit overlapping slightly. Carefully trim off the overlaps with a sharp utility knife to allow the blister to flatten with the edges of the slit butting neatly together. Once the veneer is flat, wipe off any excess glue that might have surfaced, then weight it down with heavy books overnight.

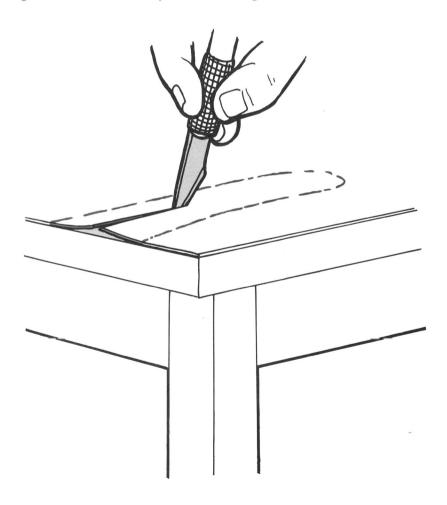

Step 5-14. Repairing a broken blister.

To repair a broken blister, you should clean out the old glue and any dirt or wax. Use a sharp utility knife and a metal straightedge to cut two diagonal slits across the grain in an X pattern.

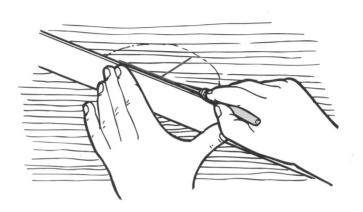

Step 5-15. Applying the glue.

Carefully lift one of the flaps and use the tip of a small screwdriver or knife to scrape the old glue from the bottom of the veneer and the top of the base wood. If the veneer is stiff or brittle, moisten it with a few drops of warm water. Scrape under the other flaps the same way. Use a drinking straw to blow out loose particles or vacuum out debris. With a toothpick, spread fresh glue under the flaps and press the veneer back into place. Wipe away the excess glue with a damp cloth.

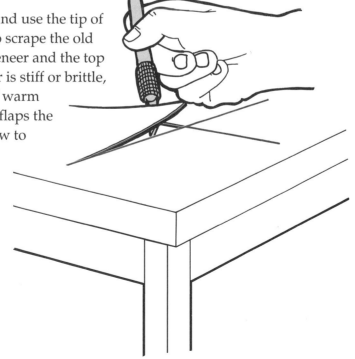

Step 5-16. Adding weight.
Place a piece of waxed paper over the repair and stack a few books over it for weight. Let dry overnight.

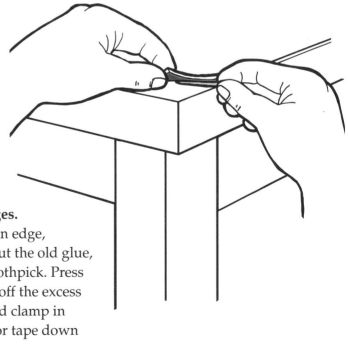

Step 5-17. Repairing loose edges.
If the veneer has separated at an edge, carefully lift the edge, scrape out the old glue, and spread new glue with a toothpick. Press the veneer back in place, wipe off the excess glue, cover with wax paper, and clamp in place with a padded C-clamp or tape down with masking tape.

Repairing Tables   73

Repairing damaged casters

Two types of casters commonly found on tables, chairs, and beds are the general-purpose caster and the ball caster. The general-purpose caster is used on light- and medium-weight furniture, such as baby cribs and beds. This caster has a metal socket that fits into a hole in the leg. A metal stem on the roller fits into the socket and allows the roller to swivel. Ball casters are usually found on furniture that might be moved about on carpet, such as small tables and some dining chairs. Ball casters are available with mounting sockets, or they can be attached to a metal plate fastened to the leg with screws.

Step 5-18.
Repairing a balky caster.
Getting to the caster might be harder than the actual repair. If the caster is on a bed, you'll probably have to take the bed apart. If it's on a chair, reaching it is not much of a problem. Once you have access to the caster, use the tip of a screwdriver to pry the roller from the socket. Spin the roller to see if it is free to turn. If not, remove it and look for any fibers, string, or other debris that might be binding the roller. If the roller is badly worn or damaged, replace it with a new one of the same size. If the roller is in good shape, spray a silicone lubricant on the stem and roller and press it back into the socket until it snaps into place.

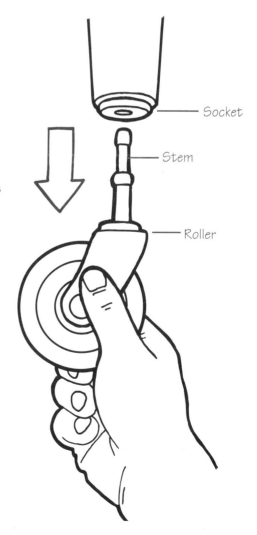

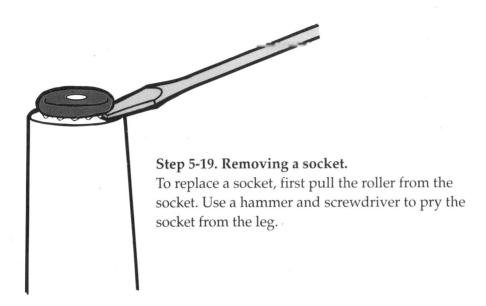

Step 5-19. Removing a socket.
To replace a socket, first pull the roller from the socket. Use a hammer and screwdriver to pry the socket from the leg.

Step 5-20. Drilling the hole.
If the leg is split, wrap it tightly with masking tape. Select a dowel and a drill bit slightly larger than the diameter of the hole and drill out the hole. Bevel the end of an appropriate-diameter dowel with sandpaper and crimp grooves with the jaws of pliers.

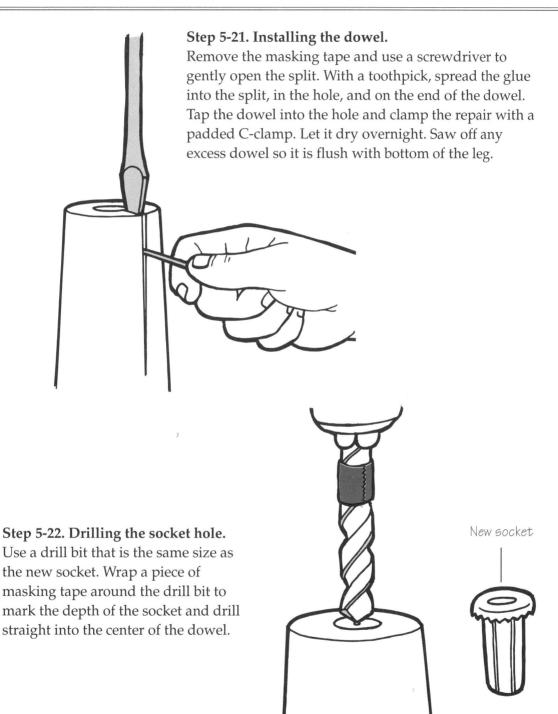

Step 5-21. Installing the dowel.
Remove the masking tape and use a screwdriver to gently open the split. With a toothpick, spread the glue into the split, in the hole, and on the end of the dowel. Tap the dowel into the hole and clamp the repair with a padded C-clamp. Let it dry overnight. Saw off any excess dowel so it is flush with bottom of the leg.

Step 5-22. Drilling the socket hole.
Use a drill bit that is the same size as the new socket. Wrap a piece of masking tape around the drill bit to mark the depth of the socket and drill straight into the center of the dowel.

New socket

Step 5-23. Installing the socket.
Press the new socket into the hole as far as you can. Now place a piece of wood over the socket and use a hammer to drive the teeth of the socket into the leg. Lubricate the stem of the roller with candle wax or silicone spray, and press the stem into the socket until the roller locks into place.

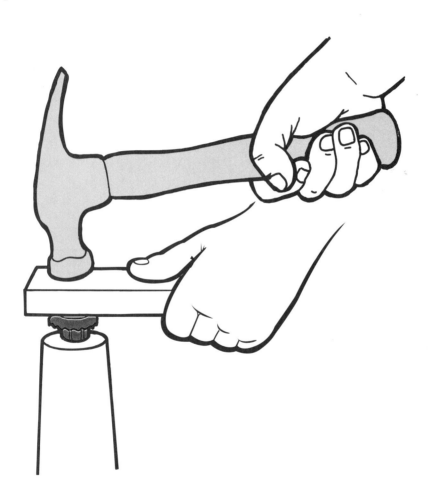

CHAPTER SIX

Drawers & Doors

Drawers and doors can stick because of loose joints, loose hinges, faulty runners, or swelling due to moisture. If a sticking drawer slides on wood runners, try removing the drawer and lubricating the runners and the edge of the drawer with candle wax or a bar of soap. If the drawer slides on metal tracks, repair any loose mounting screws, then clean the tracks with an old toothbrush and spray on a silicone lubricant.

Drawers and doors can also stick if the cabinet itself is warped. If a door sticks, first look for a loose or poorly aligned hinge and try tightening or straightening it. Repairs are usually simple, but locating the problem first is important.

Tools & materials

- Screwdriver
- Padded hammer or rubber mallet
- Wood glue
- Clamps
- Toothpicks
- Strip of canvas
- Tape measure
- Hammer, box nails, and nail set
- Nail puller
- Bar soap, candle wax, or paraffin
- Thumbtacks
- Veneer strips
- Sandpaper and sanding block
- Ruler or straightedge
- Scrap board and books or weights
- Saw (backsaw or jigsaw)
- Waxed paper
- Level
- Thin cardboard
- Utility knife
- Wood chisel

Repairing a loose joint

Examine the joint to determine the best way to take it apart. You might find a dovetail joint, a dado joint, or a rabbet joint. These joints might have nails or brads that you must remove before you separate the joint.

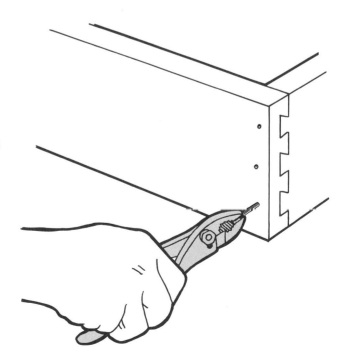

Step 6-1.
Removing the old glue.
Remove any nails or brads. Then carefully separate the joint by hand or tap it apart with a padded hammer. Try not to loosen any other joints. Use the tip of a screwdriver to scrape off the old glue.

Drawers & Doors

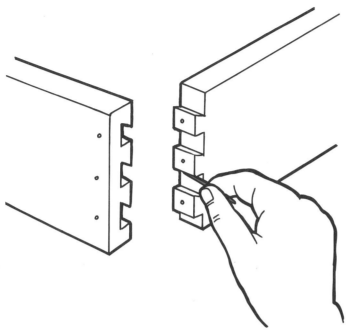

**Step 6-2.
Applying fresh glue.**
Use a toothpick to spread a thin layer of glue on all surfaces of the gluing area. Press the joint back together by hand or use a padded hammer to close the joint.

Step 6-3. Clamping.
Use a pipe clamp or a piece of rope and a wedge to hold the glued surfaces together. Wipe off the excess glue with a damp cloth. Now use a tape measure to measure from the right front corner diagonally across the drawer to the left rear corner. Repeat the step measuring from the left front corner to the right rear corner. The numbers must match if the drawer is to be square. Make any necessary adjustments in the clamp to square up the drawer. Then let the repair dry overnight.

Repairing the bottom of a drawer

Step 6-4. Removing nails.
Use a block of wood and a hammer to carefully tap inside the back of the drawer to loosen the nails holding the bottom to the back. Now turn the drawer upside down and use a nail puller, a claw hammer, or a pair of pliers and a block of wood for leverage to remove the nails.

Step 6-5. Removing the old bottom.
Slide the bottom out of the back of the drawer. If the bottom sags only a little, you can probably get by with just turning it over and reinstalling it. If it is split or otherwise damaged, you might want to install a new one. However, you can repair a split bottom by pressing the split back together and gluing a canvas strip over the split for reinforcing. If you are going to replace the bottom, take the old bottom to a hardware store that sells lumber and ask them to cut you a replacement, or you can buy the material and cut it yourself.

Step 6-6. Installing the new bottom.
Slide the bottom back into the drawer and fasten it to the back with box nails placed a couple inches apart.

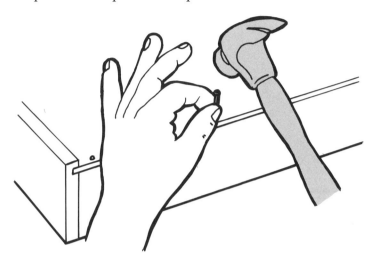

Repairing a sticking drawer with bottom runners

Step 6-7. Lubricating.
Remove the drawer and look for any protruding nail heads that might be causing the problem. Use a hammer and a nail set or large nail to drive the nail heads just below the surface. Rub a bar of soap, candle wax, or paraffin wax along the bottom of the runners and along the guides inside the cabinet.

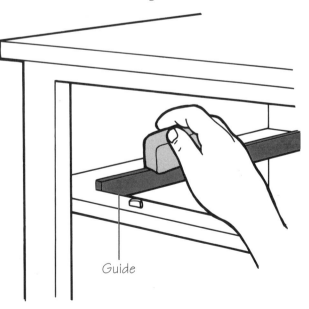

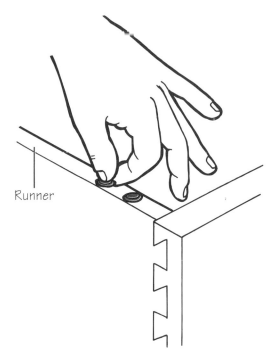

Step 6-8.
Repairing a worn runner or guide.
If the runner (on the drawer) has a curve worn in it, build it back up by pressing a few thumbtacks into the worn area. If the runner is not worn, check the guides in the cabinet. Build them back up the same way by pressing thumbtacks into the worn area. If the runner is worn evenly along the full length of the drawer, build it back up with veneer strips as shown in the following steps.

Step 6-9.
Fitting the veneer strip.
Buy a veneer strip that is long enough for both runners, if necessary. Cut it to fit the length of the runners. Sand the runner level with sandpaper fitted to a sanding block.

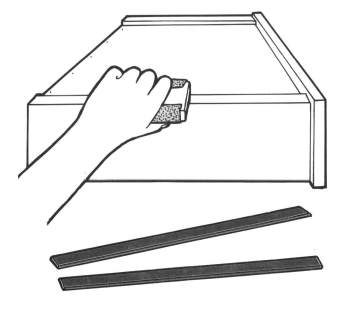

Step 6-10. Gluing the strip.
Apply a thin, even layer of glue to both the bottom of the runner and the back of the strip. Press the strip along the runner and wipe off the excess glue. Repeat the step for the other runner, if necessary. Place a board along the strip to distribute the weight and weight it down with a book or two. Allow to dry overnight, then smooth any rough edges with medium sandpaper. Add a second strip if one was not enough.

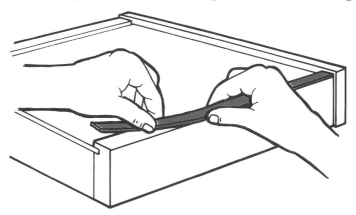

Replacing a badly worn runner

Step 6-11. Marking the cut line.
Use a ruler or straightedge to mark a line to cut out the worn area. Make the mark on the side, running the full length of the drawer but not below the bottom of the drawer. Remember that the bottom fits into grooves—you don't want to cut into them.

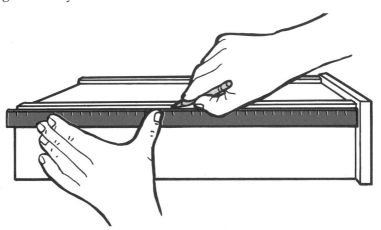

Step 6-12. Making the cut.
Before cutting out the worn part, measure the length and width of the side of the drawer and the height to your mark for the new runner. Buy new hardwood runners at a lumber supply store and have them cut or cut them yourself to these measurements. Now use a backsaw or a jigsaw with a wood-cutting blade to cut down the runner to the mark you made. If repairing both runners, cut away the other runner the same way.

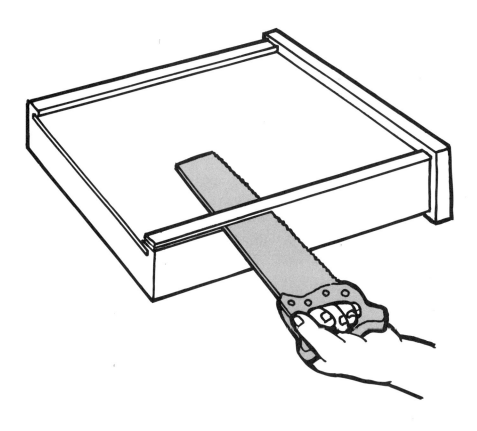

Shopping List for All Thumbs Guide to Fixing Furniture

- ☐ Adjustable pliers
- ☐ Alcohol burner
- ☐ Boiled linseed oil
- ☐ Cheesecloth
- ☐ Chisel
- ☐ Clamps
- ☐ Crosscut saw
- ☐ Denatured alcohol
- ☐ Dowels
- ☐ Drill & drill bits
- ☐ Furniture stripper
- ☐ Hammer, box nails, & nail set
- ☐ Hose clamps
- ☐ Leveling compound & rubbing felt
- ☐ Mineral spirits
- ☐ Orbital sander
- ☐ Pallet knife
- ☐ Penetrating resin & oil (tung oil, teak oil, etc.)
- ☐ Putty knife
- ☐ Respirator, rubber gloves, & safety goggles
- ☐ Rubber mallet
- ☐ Sandpaper (coarse, medium, fine, & extra fine)
- ☐ Saw (backsaw, jigsaw, or plane)
- ☐ Screwdriver
- ☐ Shellac stick
- ☐ Tack rag
- ☐ Turpentine
- ☐ Utility knife & extra blades
- ☐ Varnish
- ☐ Vise
- ☐ Wood chisel
- ☐ Wood filler
- ☐ Wood glue
- ☐ Wood putty
- ☐ Wood stain
- ☐ Wrench
- ☐ _____
- ☐ _____

Refer to the lists at the beginning of each chapter for the tools you need for individual projects.

Safety Tips

- Read carefully and follow the manufacturer's instructions for using and storing any chemicals.

- Work outdoors or in a well-ventilated room.

- Keep emergency numbers, such as the fire department and poison control center, near the phone. Make sure all family members can dial 911.

- Dispose of toxic chemicals at a hazardous waste center. Do not pour them down the drain.

- Wear safety goggles when using power saws and performing work such as chipping away dried glue.

- Wear rubber gloves when working with any caustic chemicals.

- Wear a dust respirator when sanding and work in a well-ventilated area when using toxic chemicals.

- Use power tools with grounded plugs or two-pronged plugs labeled ``double-insulated.'' Never cut off the ground prong of a plug to use in an ungrounded outlet.

- Never use any power tool in a damp location.

- Have an ABC fire extinguisher handy and know how to use it.

From *All Thumbs Guide to Fixing Furniture* by Robert W. Wood
© 1994 by TAB Books, a division of McGraw-Hill, Inc.

Drawers & Doors   85

Step 6-13.
Gluing on the new runners.
Make sure the new runners fit properly. Then apply a thin, even layer of glue to both gluing surfaces and press the runner in place. Repeat the step for the other runner.

Step 6-14. Clamping.
Cover the runner with a piece of waxed paper and place a strip of wood on top of the waxed paper and runner to distribute the pressure evenly. Repeat the steps for the other side. Now install C-clamps or weight down the runners with a couple of books placed on another board running across both runners. Wipe off the excess glue with a damp cloth. Let the repair dry overnight.

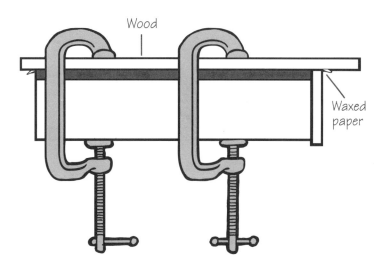

Repairing a sticking drawer with side runners

Step 6-15. Fixing protruding nails.

Some drawers have guides cut into the side of the drawer, with the runners mounted inside of the cabinet. Check the runners inside the cabinet for exposed nail heads. Use a hammer and nail set or large nail to drive the heads slightly below the surface.

Step 6-16. Lubricating runners and guides.

Use the tip of a screwdriver or knife to remove any old wax from the runners and guides. Rub a bar of soap, candle wax, or paraffin wax on the sliding surfaces of the runners and guides.

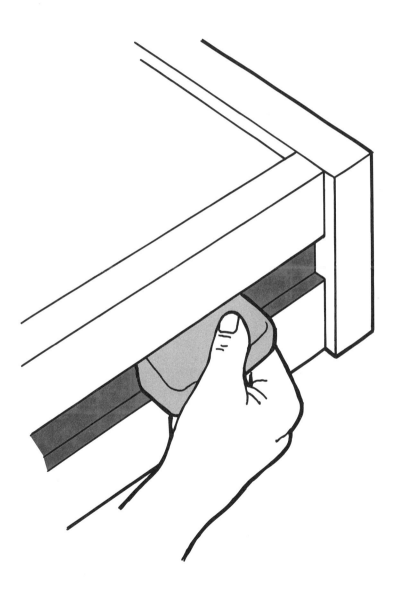

Repairing a sticking door

First make sure the cabinet is level. If not, the cabinet could be twisting and binding the door. Shim up any low corners so that the cabinet sets level on the floor. Next check for and tighten any loose hinges. If screw holes are stripped, use longer screws or plug the holes with a couple toothpicks dipped in glue and reinstall the screws. If the door still sticks, examine the hinges to be sure that the leaves fit flush with the frame. If the recessed area (the mortise) is too deep, you can install a shim. If it is too shallow, you can cut it deeper. If the door still sticks, try sanding off the door to relieve the bind.

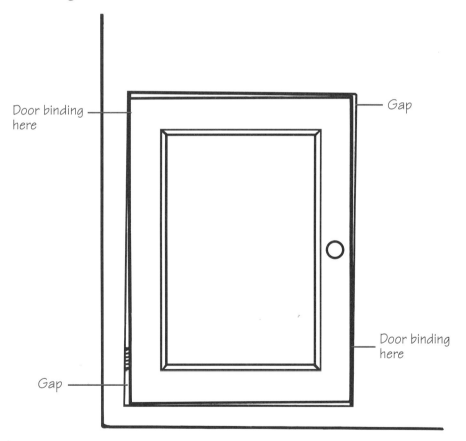

In the illustration, the door is binding at the upper left and the lower right corners. To repair, shim the top hinge or cut the mortise of the bottom hinge deeper. If the opposite corners bind, repair the opposite hinges.

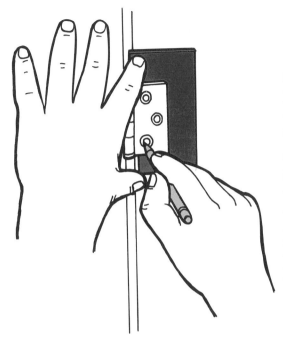

Step 6-17. Cutting a cardboard shim.
Prop up the bottom of the door so that it won't sag; then remove the screws from the part of the hinge fastened to the cabinet. Slip a thin piece of cardboard (cut from a cereal box, cracker box, or the like) behind the hinge leaf and trace the outline and screw holes with a pencil. Use heavy scissors or a utility knife to cut out the shim; use a nail to punch holes for the screws.

Step 6-18. Installing the shim.
Place the shim in the recessed area behind the hinge leaf. Press the hinge leaf in place to see if it fits flush with the surface. If not, add another shim. Reinstall the screws and see if the door still sticks.

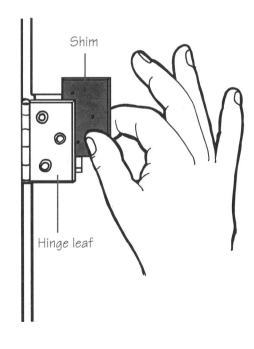

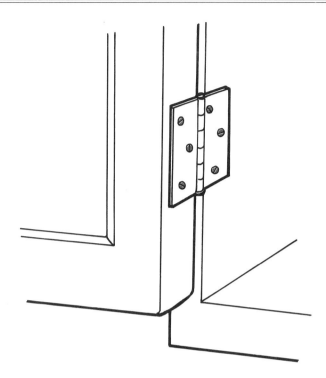

**Step 6-19.
Cutting a deeper mortise.**
Examine the hinge that should be set deeper and notice how much deeper it should be. Remove the door from the cabinet by unscrewing the leaves from the cabinet frame.

Step 6-20. Cutting the outline.
Use a utility knife to cut along the outline of the mortise to about the depth you need to cut with a chisel.

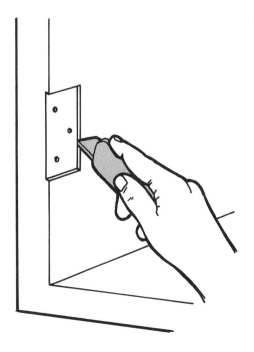

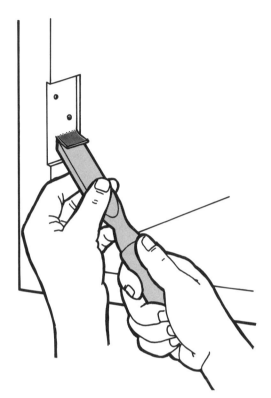

Step 6-21. Using the chisel.
Hold the chisel with the beveled side down, facing the wood, and chisel off a thin layer of wood. Apply smooth, even pressure to the chisel, making sure the recessed area is flat. Remount the door and try it for fit.

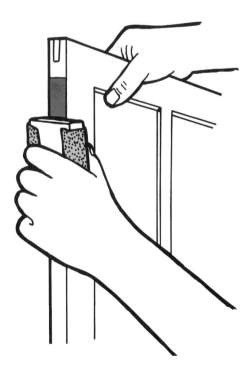

Step 6-22. Sanding a door to fit.
If the door still sticks, look for worn places and rub marks along the edge of the door to see where it has been binding. Use medium sandpaper and a sanding block to cut down these places slightly—just enough to free the door. You don't want to take off too much.

CHAPTER SEVEN

Furniture Touch-up

Most surface flaws can be corrected without stripping and refinishing the entire piece. Cigarette burns, marks made by hot and cold drinks, dents from falling objects, and accidental scratches are common injuries to furniture. Even accumulated wax and grime can make a finish dull or sticky.

Some scratches can be filled by rubbing a crayon into it. Shallow scratches can often be hidden by dabbing with a cotton swab dipped in iodine or liquid or wax shoe polish. Small dents can often be removed with nothing more than a damp cloth and an electric iron. Surfaces can be revived with a mixture of turpentine and boiled linseed oil. (Boiled linseed oil doesn't mean you heat it and bring it to a boil; it means the manufacturer has treated it with driers so it will dry.)

For filling gouges, you might have to turn to a professional product called a shellac stick. It comes in assorted colors to match your finish. Working with a shellac stick can be tricky. It has to be heated, preferably by an alcohol burner. The stick is melted into the gouge and spread with a pallet knife or an old kitchen knife.

You can save the cost and mess of refinishing by doing touch-up if the furniture has only slight surface damage. Always wear rubber gloves and work in a well-ventilated area when using toxic, caustic, or flammable chemicals.

Tools & materials

- Pocket knife
- Fine sandpaper
- Clean cloth
- Old paintbrush
- Kitchen knife
- Wax stick or crayon
- Mineral spirits
- Shellac stick
- Alcohol burner and denatured alcohol
- Pallet knife
- Fine steel wool and sanding block
- Leveling compound and rubbing felt
- Blotter
- Iron
- Household ammonia
- Mineral oil, cooking oil, or olive oil
- Wood putty
- Putty knife
- Boiled linseed oil
- Turpentine
- Glass jar with lid
- Lacquer thinner

Repairing a cigarette burn

Step 7-1. Removing the charred wood.

Use a sharp knife with a curved blade—a pocket knife will do—to scrape away the charred wood. Try not to damage the surrounding area, and stop scraping when you reach solid wood. Smooth the scraped area with fine sandpaper and brush away the loose debris with a clean cloth or an old paintbrush.

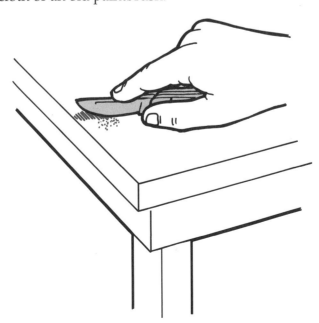

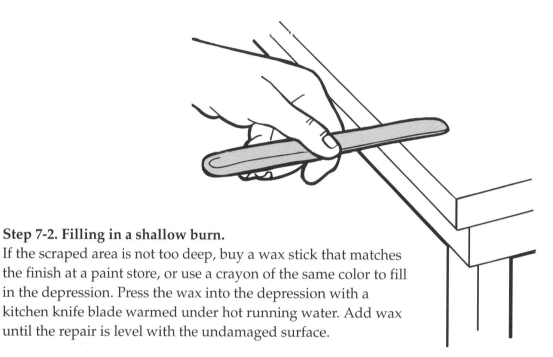

Step 7-2. Filling in a shallow burn.
If the scraped area is not too deep, buy a wax stick that matches the finish at a paint store, or use a crayon of the same color to fill in the depression. Press the wax into the depression with a kitchen knife blade warmed under hot running water. Add wax until the repair is level with the undamaged surface.

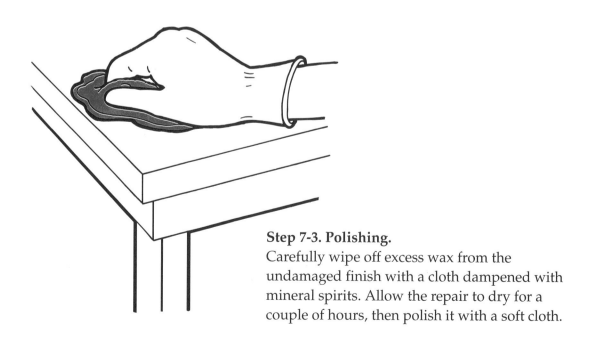

Step 7-3. Polishing.
Carefully wipe off excess wax from the undamaged finish with a cloth dampened with mineral spirits. Allow the repair to dry for a couple of hours, then polish it with a soft cloth.

Using a shellac stick

A shellac stick is a better way to repair deeper problems such as burns, gouges, and cracks, but its use is a bit more complicated. The shellac makes a hard patch that is difficult to remove if you make a mistake, while a wax patch can easily be repaired. You might want to practice on a piece of scrap before you try the actual repair.

Step 7-4. Gathering the materials.
Buy a shellac stick that matches the color and gloss of your furniture's finish. You also need an alcohol burner, denatured alcohol, and a pallet knife or grapefruit knife. You should find the supplies at a good paint store or a large hardware store.

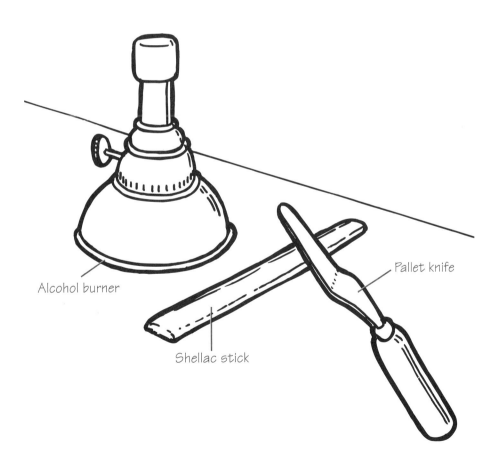

Step 7-5. Scraping the damaged area.
Use the tip of a knife to carefully scrape the damaged area down to good wood.

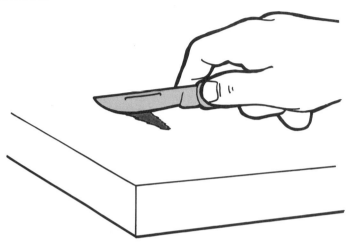

Step 7-6. Applying the shellac.
Light the alcohol burner and heat the blade of the pallet knife for several seconds. Place the end of the shellac stick over the damaged area and press the heated blade against the end of the stick. The shellac will start to melt. Allow the shellac to drip into the damaged area and smooth it with the knife. You'll have to reheat the blade to keep the shellac dripping and to work it into the repair. Build up the damaged area just a little above the surrounding surface, then smooth it with the heated knife. Now let the repair cool and harden.

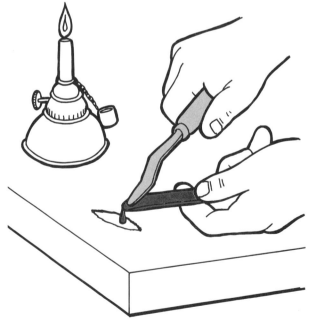

Step 7-7. Cutting down and smoothing.
Use fine steel wool (labeled 3/0 and called three-ought or finer) wrapped around a flat block of wood to smooth the repair flush with the surrounding surface. Rub with the grain of the wood. You also can use leveling compound and a block of rubbing felt (available at paint stores) to smooth the surface. Apply furniture wax and polish with a soft cloth to bring out the luster.

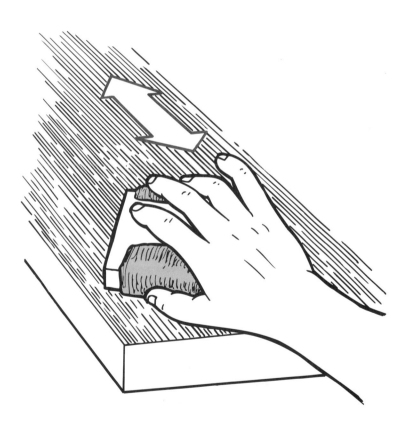

Removing white rings or spots

White rings can be caused by condensation on a cold drink, alcohol from spilled shaving lotion or perfume, or even by a hot cup of coffee.

Step 7-8. Removing a water spot.
First try placing a blotter over the spot and pressing with a warm iron. If this method doesn't work after a couple of tries, you'll have to rub the spot with a fine abrasive and some type of lubricant (Step 7-10).

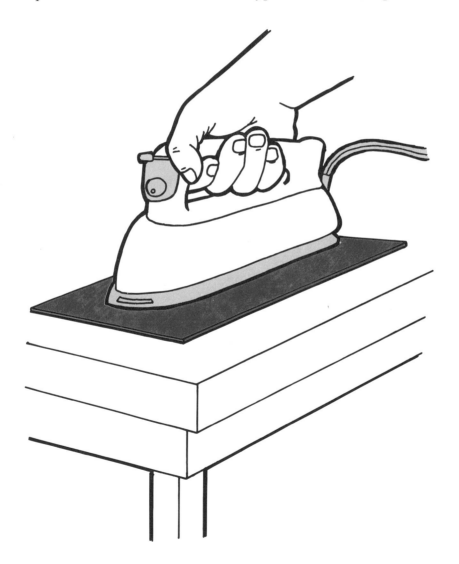

Step 7-9. Removing an alcohol spot.
If you're fast enough after the spill, try very lightly wiping the spot with a cloth moistened with a little household ammonia. If this method doesn't work, rub with a fine abrasive and lubricant (Step 7-10).

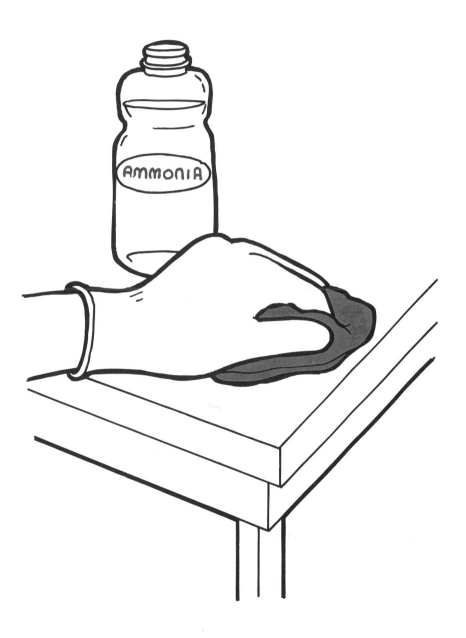

Step 7-10. Rubbing with an abrasive and lubricant.
Dampen a piece of fine steel wool (grade 3/0) with mineral oil and lightly rub the spot with the grain. Wipe off the excess oil with a dry cloth. If steel wool is not handy, use a cloth and scouring powder, baking soda mixed into a paste with petroleum jelly, cooking oil, olive oil, or even mayonnaise. Wipe off the excess with a damp cloth, then dry with another cloth. Apply furniture polish or oil to remove any dull places.

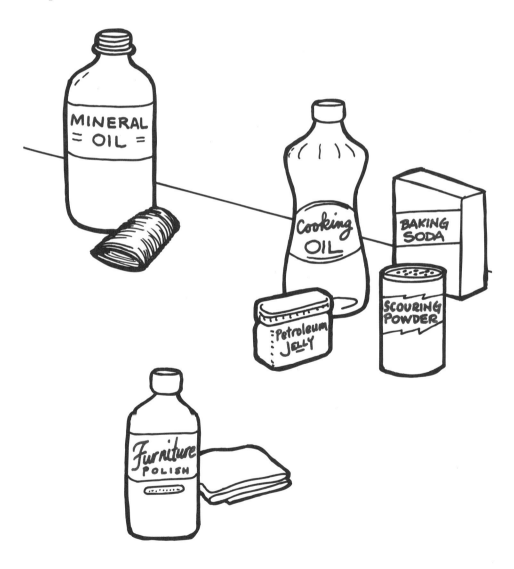

Repairing a deep gouge

Step 7-11. Filling with wood putty.

Use the tip of a knife to clean out the damaged area. Buy a water-based wood putty that matches the color of the finish, or add stain to the putty. Use a putty knife to firmly press the putty into the cavity. If the gouge is not too visible, overfill the repair slightly and let dry. Then sand smooth with medium sandpaper, sanding with the grain. If the damage is in a conspicuous place, don't completely fill the cavity, but apply a layer of stick shellac that matches the finish (Steps 7-4 through 7-7).

Step 7-12. Leveling the patch.
Rub down the repair with 3/0 steel wool wrapped around a flat block of wood, or a felt block and leveling compound. Wipe off the patch with a clean cloth.

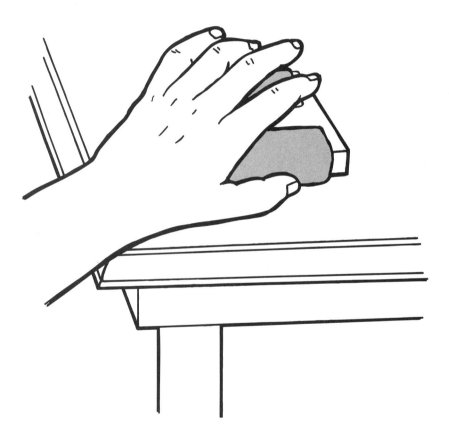

Reviving a finish

The quickest and easiest way to revive a finish is to use one of the prepared products available at paint stores. These liquid refinishers can revive varnished, lacquered, or shellac finishes, but the better brands are expensive. Just follow the manufacturer's instructions and you should get good results. To try other methods, use the following steps.

Step 7-13. Cleaning the surface.
Use a cloth dampened with mineral spirits to loosen wax buildup. Clean small areas at a time and wipe off residue with a dry cloth.

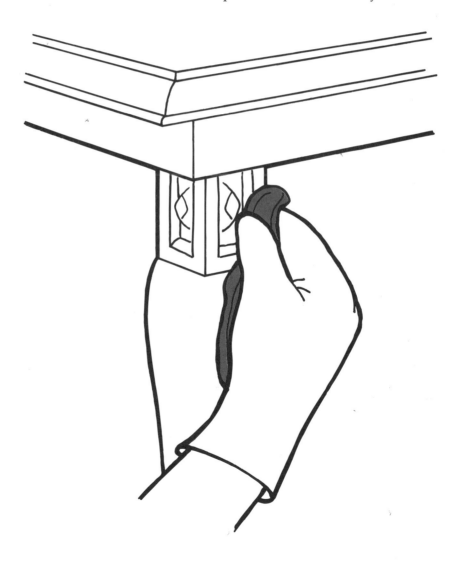

Step 7-14. Reviving a varnished surface.
Mix one part boiled linseed oil with two parts turpentine in a glass jar with a lid. Shake the jar to mix the solution. Use a 3/0 steel wool dipped in the mixture to lightly go over the surface. Make the strokes with the grain. Then use a lint-free cloth, wiping with the grain, until the surface has a thin, oily film. Let dry and repeat the steps if necessary. If the finish is still dull, apply furniture wax and buff with a soft cloth.

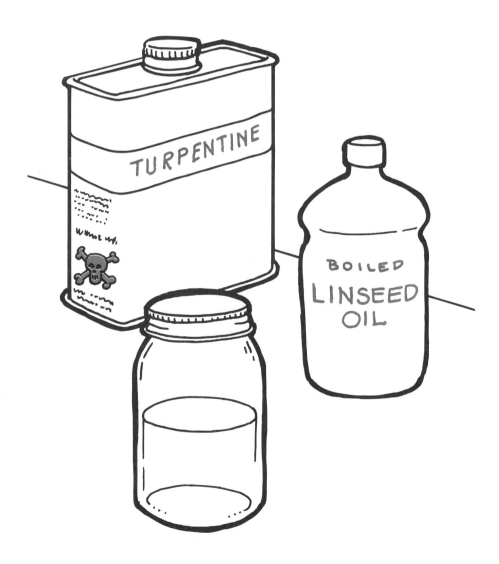

Step 7-15. Reviving a shellac or lacquer finish.

Denatured alcohol dissolves dried shellac, but it can cause a white film on the surface if it is applied in damp weather. If the finish is dull or has minor scratches, wipe the surface with a cloth dipped in the alcohol. If the finish is cracked, use a varnish brush to apply light strokes with the grain until the cracks have disappeared. After the finish has dried, smooth it with fine steel wool and apply a coat of wax. To revive a lacquered finish, use the same steps, but use lacquer thinner instead of denatured alcohol.

CHAPTER EIGHT

Preparing the Surface for the Finish

If the old finish could not be revived, you'll have to strip it and apply a new finish. It is a messy job, and the best removers are the most expensive. Most chemical strippers pose some type of health or fire risk, so the easiest way around this problem is to have a professional do the job for you. However, for the determined do-it-yourselfer, good-quality strippers are available in paste or liquid form.

Some strippers are water-based and can be washed off with water, but water can raise the wood fibers, requiring extra sanding. The best removers contain methylene chloride, a powerful solvent that works well on a variety of finishes. A paste stripper works better than a liquid on vertical surfaces.

Try to do the job outdoors and out of direct sunlight. At least work in a well-ventilated area. Have plenty of old newspapers handy, as well as a good supply of clean lint-free cloths, and follow the manufacturer's instructions. Safely store or dispose of used rags and supplies.

Tools & materials

- Sheet of plastic
- Newspapers
- Old pans
- Furniture stripper
- Old paintbrush
- Putty knife
- Old toothbrush, cotton swabs, or toothpicks
- Clean cloths
- Denatured alcohol
- Wide-mouth jar
- Fine steel wool
- Clothing spot remover
- Sandpaper and sanding blocks
- Tack rag
- Turpentine
- Varnish
- Liquid laundry bleach
- Wood filler
- Piece of burlap
- Wood stain
- Shellac
- Cheesecloth

Step 8-1. Applying the stripper.
Use a table or other surface that will put the furniture at a comfortable height for you to work on. Cover the table with a sheet of plastic for protection, then several newspapers. Place old metal pans under the legs to catch the excess stripper. Start on the largest horizontal surface and, working in the same direction, use an old paintbrush to pat the stripper evenly over the surface. Don't try to work the stripper into the finish—let the stripper do the work.

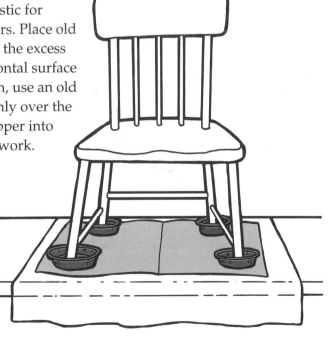

Step 8-2. Removing the old finish.
After about 5 minutes, the old finish will begin to bubble and lift from the surface. Wait until it stops bubbling, then use a wide putty knife to scrape off the old finish and stripper. Work with the grain and try not to gouge the wood. Wipe off the putty knife with newspaper. Repeat the steps until you have reached bare wood. Turn the furniture to the next largest surface and repeat the steps. Then strip the smaller parts. Use an old toothbrush, cotton swabs, or toothpicks on any intricate carvings.

Step 8-3. Cleaning the wood.
After you have stripped the finish down to bare wood, clean off any stripper or finish residue. Normally the stripper manufacturer recommends a solvent. If not, pour a little denatured alcohol into a wide-mouth jar and use a soft cloth dipped into the alcohol to rub off any residue. Remove any stubborn residue with fine steel wool dipped in the alcohol. Always work with the grain.

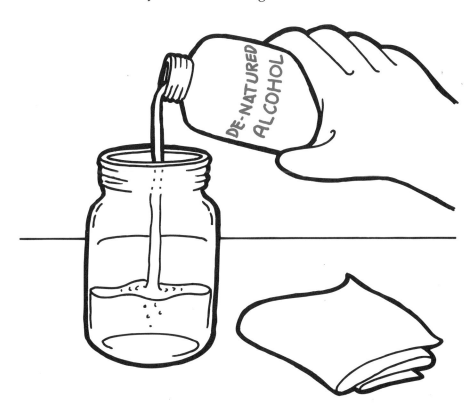

Preparing the wood

Check closely for any surface blemishes and repair them now. Remove any grease spots with clothing spot remover. You should now be ready to smooth the wood.

Step 8-4. Smoothing with sandpaper.
Begin smoothing with medium sandpaper. Use sanding blocks and sandpaper, always sanding with the grain. When sanding turned surfaces such as legs and backs of chairs, use a narrow strip of fine sandpaper and sand across the grain of the turning. Frequently tap the sanding block against the corner of the work table to knock the sawdust from the paper. Use a curved sanding block or a dowel for smoothing curves. Repeat the sanding using fine, then very fine, sandpaper. When you have a smooth surface, brush off the dust with an old paintbrush and go over the entire surface with a cloth dampened with mineral spirits or turpentine or a tack rag.

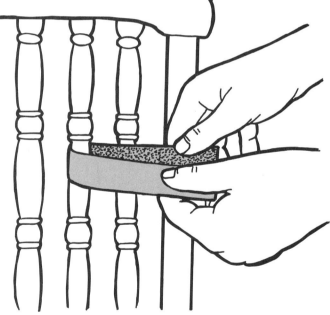

Step 8-5. Making a tack rag.

If a cloth dampened with mineral spirits or turpentine didn't work well enough, buy a tack rag at a paint store or make your own. Use cheesecloth or well-laundered lint-free cotton cloth. First dip the cloth in warm water and wring it out. Then saturate it in turpentine and wring it out a second time. Next dip it in varnish and fold and squeeze the cloth until the varnish is spread evenly over the cloth. Squeeze any excess varnish from the cloth. It should be sticky enough to pick up dust particles without leaving any streaks of varnish. You now have a tack rag. Store it in a sealed glass jar when not in use. If the rag dries out, restore it with a few drops of water and turpentine.

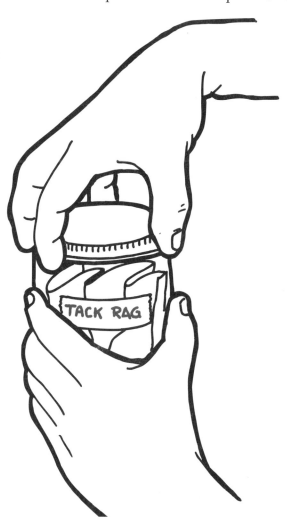

Preparing to stain

If you are not staining, the next step depends on the type of wood you have and the type of finish you want. You might want to bleach the wood for a light finish. If you want a natural finish, you'll leave the grain open and apply a penetrating oil. If the grain is open, but you want a lacquer or varnish finish, apply a filler first to fill the grain. If the wood is close-grained, you can apply the finish.

Step 8-6. Bleaching.
After the furniture has been completely cleaned, all of the old finish is gone, and the piece is free of all dirt, grease, and wax, rub lightly over the surface with fine steel wool. Now pour a small amount of liquid laundry bleach into a wide mouth jar. Use a paintbrush with nylon bristles or a rag to spread the bleach evenly, following the grain. The wood should start getting lighter almost immediately. When the wood is light enough, rinse it off with plain water and dry it with old rags. Let the piece dry completely. When it is dry, you will see that the wood fibers have raised. Smooth the surface with extra-fine sandpaper or 3/0 or 4/0 steel wool. Remove dust particles with an old paintbrush and a tack rag. You can now apply a stain, sealer, or the finish.

Step 8-7. Applying a filler.

Fillers are applied to close the pores of coarse-grained wood to produce a smooth surface for the finish coat. Fillers should not be used if you are going to apply a penetrating finish such as tung oil or teak oil. Buy a paste filler (one with a silex powder base is best) that is the same color or slightly darker than the wood. Thin it, according to the instructions on the label, to the consistency of thick cream. Use a stiff brush to brush the filler with the grain, then across the grain, working it into the wood.

Step 8-8. Removing excess filler.

Let the filler set up for 5 or 10 minutes, then rub a piece of folded burlap across the grain to remove the excess filler from the surface. Now use a soft cloth, rubbing with the grain, to remove any remaining filler. Remove filler from carved areas with an old toothbrush. Allow the filler to dry a day or two.

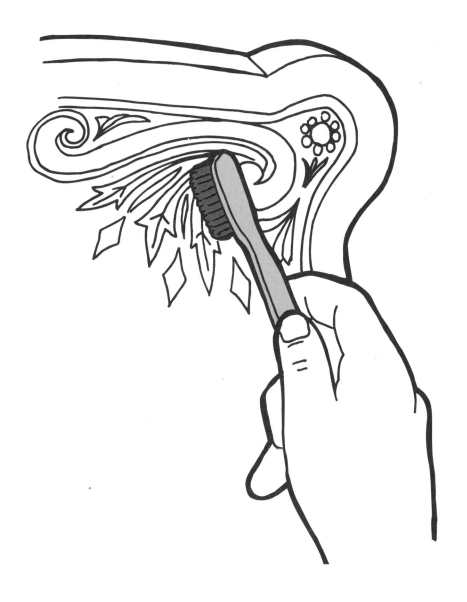

Staining

Stains soak into the wood, add color, and still let the grain show. The color of the stain varies depending on the type of wood. Your paint store should have a chart showing the effect of the stain on different types of wood. Refer to the chart for your selection. Stains are available with water, alcohol, and oil bases. The oil-based stain is the best for the home refinisher. It is inexpensive and easy to apply. Try to find a ready-mixed stain that suits you. These types are already diluted and need only frequent stirring as you work. Select a shade slightly lighter than the one you want, as stains tend to darken when they are dry and under the finish.

Step 8-9. Applying the stain.

The wood should be clean, smooth, and free of dust. Check the end-grain areas, such as the edge of a chair seat. The wood there is more absorbent and will be darker. Apply a thinned coat of shellac to seal the pores. Stir the stain well, then pour a small amount in a wide-mouth jar. Use a pad of folded cheesecloth or a paintbrush to apply the stain. Try to avoid dripping. Test the stain in an obscure place to see if you like the shade. Start at the least conspicuous place—the back, the legs, then the top. If possible, turn the piece as you work so that you are always applying the stain to a horizontal surface. If you must apply stain to a vertical surface, work from the top down so that you wipe the excess away as you go. Apply a thin, even coat with long, smooth strokes, following the grain. Use a dry cloth to wipe off excess stain as you go.

CHAPTER NINE

Finishing

The finish is the coating you apply to the wood to protect the surface and bring out the beauty of the furniture. It can be a penetrating finish or a surface finish. A penetrating finish might use linseed oil or tung oil. It is worked into the wood to give it a natural look. Other penetrating finishes, made from resin, combine the stain, sealer, and finish coat in one. They seal, preserve the wood, produce a rich finish, and are easy to apply. A surface finish uses something like varnish or lacquer and requires several coats with smoothing in between.

Tools & materials

- Penetrating oil (tung oil, teak oil, etc.)
- Turpentine
- Boiled linseed oil
- Clean cloths
- Penetrating resin
- Rag or brush
- Varnish
- Mineral spirits
- Shellac
- Denatured alcohol
- Fine steel wool
- Clean 2-inch brush
- Lacquer (spray)

Step 9-1. Applying a penetrating oil finish.
You can buy prepared penetrating oil or you can make your own. To make your own, mix tung oil or boiled linseed oil with an equal amount of turpentine in a wide-mouth jar. Wipe the surface of the furniture with a tack rag. Dip a soft cloth into the oil mixture and spread it generously over the surface, working with the grain. Wait a few minutes for the oil to soak in, then apply more oil to any dry places. Continue applying the oil until the wood no longer absorbs it. Now use a fresh cloth to wipe away the surface oil, and polish vigorously with a soft cotton cloth.

Step 9-2. Applying a penetrating resin finish.

Make sure all of the old finish has been removed and the surface is free of dust. Read and follow the manufacturer's instructions; applications can vary from one brand to another. If possible, turn the furniture so that the part you are working on is horizontal. Use a rag or brush to apply a generous coat so that it soaks in. Keep the surface wet and let the resin soak in for the time recommended by the manufacturer. Apply more resin to any dull spots that appear. When the penetration time is up, wipe off excess resin with a cloth, then wait for the recommended time and apply a second coat. When the last coat is dry, apply a good furniture wax for a mirror finish.

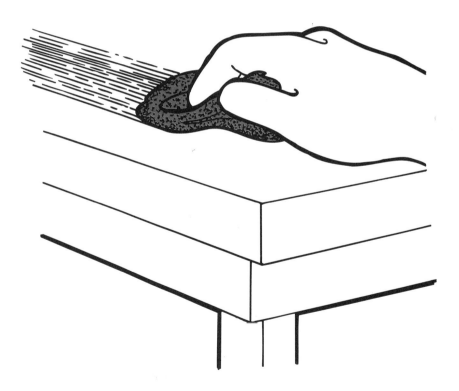

Applying varnish

Step 9-3. Preparing the surface.

The surface of the wood must be clean and dry. Try to have the room temperature between 70 and 80 degrees, with plenty of light. For best results, apply a sealer coat of shellac thinned with an equal part of denatured alcohol. Let dry overnight, then smooth the surface with 3/0 steel wool. Remove all traces of dust. For a thick, built-up finish, apply two or more coats of shellac before applying the varnish. Furniture varnish or floor varnish works, just don't use spar varnish. Buy varnish in a small can—a half pint is probably enough. Pour the varnish into another can and add about an ounce and a half of mineral spirits. Stir the mixture.

120 Fixing Furniture

Step 9-4. Applying the varnish.
Apply the varnish with a clean 2-inch brush. A cheap brush is as good as an expensive one. Work with the light reflecting across the surface so you can see skips. Wet the whole surface you are working on, using hard strokes to spread the varnish onto the surface. When the complete surface is covered with a well-brushed, thin coat, hold the brush at a 45-degree angle and lightly brush once across the grain, then once with the grain. Then don't touch it. Any brush marks you see will probably level themselves out. If you mess with it, they won't. Now let the varnish dry for a couple days. Use 4/0 steel wool to smooth the surface, and wipe off the dust before applying the second coat, if necessary. One coat is usually enough.

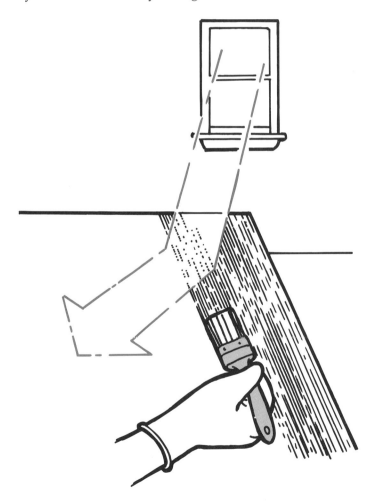

Step 9-5. Applying shellac.

Shellac makes a good sealer. It is easy to apply, but it can be damaged by water, heat, or alcohol. It is not a good finish for table tops. Shellac is sold in *cuts* (the amount of shellac resin dissolved in alcohol). Pour the shellac (3- or 4-pound cut) into a can and add denatured alcohol. Mix one part shellac to three parts alcohol. Apply the shellac with a paintbrush as quickly as possible. Smooth the shellac with a few quick strokes, no more. Let dry about three hours, smooth the surface with 3/0 steel wool, wipe off the dust, and apply a second coat. Let the second coat dry overnight, rub down with 3/0 steel wool, and wax the finish.

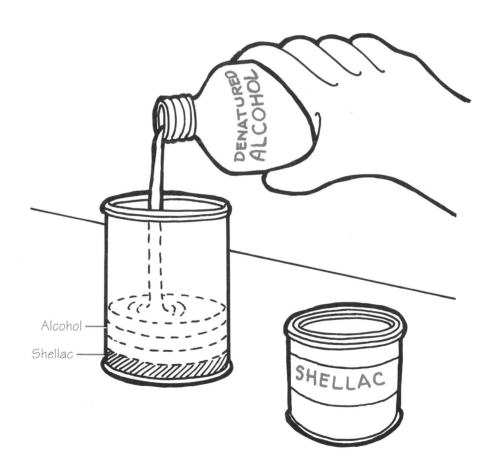

Step 9-6. Applying lacquer.
Lacquers produce hard, brilliant finishes, but they are difficult to work with because they dry so fast. For larger jobs, lacquering is best left to a professional. Lacquer is highly flammable, so work outdoors out of direct sunlight. For small jobs, use a spray can. After the wood has been sealed, smoothed, and dusted, shake the can for a couple minutes. Hold the can about a foot from a vertical surface, and spray in a smooth, steady sweep. Shake frequently between passes. Spray first in one direction, then spray on the return pass, overlapping half of the first pass. When the first coat is finished, let it dry about a half hour. Gently smooth the surface with fine steel wool; then wipe clean with a tack rag before applying the second coat. You'll need to apply several coats. Let the final coat dry a day or so, then smooth and polish with 4/0 steel wool and furniture oil.

Glossary

ABC fire extinguisher A dry-chemical fire extinguisher suitable for putting out paper, electrical, and oil fires.

apron A wood frame used to support a table top.

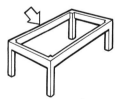

boiled linseed oil Linseed oil that has been pretreated to dry faster than regular linseed oil.

124 Fixing Furniture

caustic chemicals Chemicals that burn or destroy tissue by chemical action. Corrosive.

dado A rectangular groove cut into the side of a board so that another board can be fitted into it.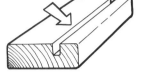

dovetail joint A wedge-shaped part that fits into a wedge-shaped opening to form an interlocking joint.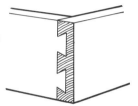

dowel A wooden peg or pin.

dowel joint A wooden pin fitted into a hole to form a joint.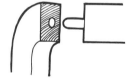

dowel socket The hole a dowel fits into to form a joint.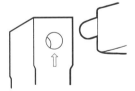

filler A preparation used to fill the open grain in wood before painting or varnishing.

Glossary **125**

grain The arrangement or direction of the layers of wood.

lacquer A resinous coating dissolved in solvent that evaporates rapidly on application.

mineral oil An oil derived from petroleum.

mineral spirits A thinner with an oil base.

mortise A hole or recess cut into wood to receive a projection.

mortise-and-tenon joint A joint where a projection (tenon) fits into an opening (**mortise**).

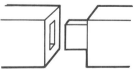

rabbet joint A groove or recess cut into the edge of a board so that another board fits into it.

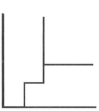

solvent A substance, normally liquid, that dissolves or can dissolve another substance.

stiles The vertical pieces in the back frame of a chair.

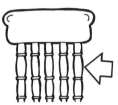

wing nut A nut with flared sides that allow turning with the thumb and forefinger.

Index

A
ABC fire extinguisher, 7
alcohol spot, 99

B
backsaw, 10
band clamp, 13
bar clamp, 12, 36
belt clamp, *see* band clamp, 13
bentwood chairs, 56-57
bleaching, 112
bolted joint, 62-65

C
C-clamp, 11, 34
casters, repair, 73-76
chair back, 46-48, 51-53
  loose, 51-53
  regluing, 46-48
chair legs, 44-45, 49-50
  loose, 49-50
  regluing, 44-45
chair repair, 38-57
  back loose, 51-53
  back regluing, 46-48
  bentwood chairs, 56-57
  foot rails, 41-43

frame chairs, 49-55
joints, 40
leg regluing, 44-45
legs loose, 49-50
platform chair, 41-48
tools and materials, 41
types, 39-40
wobbly chairs, 54-55
chairs, 39, 56-57
  bentwood, 56-57
  frame, 39
  platform, 39
chemicals, 3-4
  disposing of, 4
  storing, 3
chisel, 9, 31-32
  techniques, 31-32
  wood, 9
cigarette burn, touch-up, 93-97
clamps
  band, 13
  bar, 12, 36
  belt, 13
  C-clamp, 34
  hand screw, 35
  improvised, 37
  pipe, 12, 36
  techniques, 33-38

  twine, 34
  web, 35
  wood, 11
corner brace, 58
crosscut saw, 11, 27-29
  cut, completing, 29
  cut, starting, 28
  saw preparation, 27
  techniques, 27-29

D
denatured alcohol, 109
deep gouge, repair of, 101-102
doors, 88-91
  leveling, 88
  mortise cutting, 90-91
  sanding to fit, 91
  shimming, 88
  repairing, 88-91
  sticking, 88-91
dovetail joints, 78
dowel joint, 40
  broken, 22-25
    dowel, cutting, 24
    dry run, 24
    gluing, 25
    joint aligning, 23

dowel joint *continued*
  old dowel, removing, 22
  socket holes, aligning, 22
  sockets, drilling, 23
 regluing, 17-21
  clamp, applying, 21
  dowel, sanding, 18
  dowel socket, sanding, 19
  dry run, 19
  glue, applying, 20-21
  wedge use, 20
drawer repair, 78-87
 bottom repair, 80-81
 loose joint, 78-79
 runner replacement, 83-87
  bottom, 83-85
  side, 86-87
 sticking, 81-83
  lubricating runner, 81
  worn runner or guide, 82
 veneer strip, adding, 82-83
drawers and doors, 77-91
 doors, 88-91
 drawers, 78-87
 tools and materials, 78
drill, electric, 9
drop-leaf table, 59
dust mask, 13
dust respirator, 6

E
electricity, safety, 6-7
emergency phone numbers, 2
extension table, 59
 repair, 66-76
 runners, wood, 66-68
 veneer, regluing, 69-72

F
filler, 113-114
 application, 113
 excess removal, 114
finish reviving, 103-105
 cleaning surface, 103
 shellac or lacquer finish, 105

varnished surface, 104
finishing, 106-122
 lacquer application, 122
 oil finish, 117
 resin finish, 118
 shellac application, 121
 stain preparation, 112-114
 staining, 115
 surface preparation, 106-115
 tools and materials, 116
 varnish application, 119-120
 wood preparing, 110-111
fire extinguisher, ABC, 7
foot rails, loose, 41
frame chairs, 39, 49-55
 loose back, 51-53
 loose legs, 49-50
furniture touch-up, 92-105

G
glossary, 123-126
gloves, rubber, 5, 14
glue, woodworker's, 13
goggles, 5, 13
gouge, 101-102
 deep, 101-102
 wood putty, 101-102

H
hand screw clamps, 35
hinge beds, split, 25-26
hinges, 25-26
 loose screws, 26
 repairing, 25-26
 split hinge beds, 25-26

J
joints
 dovetail, 78
 dowel, 17-21, 40
 mortise-and-tenon, 40

K
knife, utility, 12

L
lacquer, 105, 122
 application, 122
 finish, reviving, 105

M
mallet, rubber, 10
mask, dust, 13
masking tape, 9
mortise-and-tenon joint, 40
mortise, 90-91

O
oil finish, penetrating, making, 117
orbital sander, 10, 30-31
 techniques, 30-31

P
penetrating oil finish, making, 117
penetrating resin finish, 118
phone numbers, emergency, 2
pipe clamp, 12, 36
platform chair, 39, 41-48
 back regluing, 46-48
 foot rails, loose, 41-43
 legs, regluing, 44-45
 repair, 41-48

R
resin finish, penetrating, 118
respirator, 6, 14
 dust, 6
rubber gloves, 5, 14
rubber mallet, 10
runner, 83-87
 bottom, 83-85
 lubricating, 81
 replacement,
  bottom, 83-85
  side, 86-87
 worn, 82

S
safety, 1-7

Index

chemicals, 3-4
  disposing of, 4
  storing, 3
electricity, 6-7
fire extinguisher, ABC, 7
gloves, rubber, 5
goggles, 5
labels, reading of, 4
phone numbers,
  emergency, 2
respirator, dust, 6
sander, orbital, 10
sandpaper, 14
saw, crosscut, 11
screwdriver, 9
shellac application, 121
shellac finish, reviving, 105
shellac stick, 95-97
shim, cardboard, 88-89
split hinge beds, 25-26
stain preparation, 112-114
  bleaching, 112
  filler, 113-114
    application, 113
    excess removal, 114
staining, 115
steel wool, 14-15
sticking drawers, 81-83
  lubricating runner, 81
stripping surface, 107-108
  wood cleaning, 109
surface preparation, 106-115
  finishing, 106-115
  stripping, 107-108
  tools and materials, 107

T

table apron, 58
table joint, bolted, 62-65
table repair, 58-76
  apron, 58
  casters, damaged, 73-76
  corner brace, 58
  drop-leaf, 59
  extension table, 66-76
  tools and materials, 60
  types, 59
  wood joints, loose, 60-62
table veneer, regluing, 69-72
tack rag, 111
tape, masking, 9
techniques, 16-37
  chisel, 31-32
  clamps, 33-38
  crosscut saw, 27-29
  dowel joint, 17-25
    broken, 22-25
    regluing, 17-21
  hinges, repairing, 25-26
  orbital sander, 30-31
  tools and materials, 16
tools, 8-15
  backsaw, 10
  band clamp, 13
  bar clamp, 12
  C-clamp, 11
  crosscut saw, 11
  drill, electric, 9
  goggles, safety, 13
  mask, dust, 13
  orbital sander, 10
  pipe clamp, 12
  respirator, 14
  rubber gloves, 14
  rubber mallet, 10
  sandpaper, 14
  screwdriver, 9
  steel wool, 14-15
  tape, masking, 9
  utility knife, 12
  wood chisel, 9
  wood clamp, 11
  woodworker's glue, 13
touch-up, 92-105
  alcohol spot, 99
  cigarette burn, 93-97
  finish reviving, 103-105
  gouge, deep, 101-102
  shellac stick, 95-97
  tools and materials, 92
  water spots, 98
  white rings or spots, 98-100

U

utility knife, 12

V

varnish application, 119-120
varnished surface, reviving, 104
veneer, 69-72, 82-83
  adding weight, 72
  broken blister repair, 71
  glue application, 71
  large blister repair, 70
  loose edges, 72
  regluing, 69-72
  small blister repair, 69
  strip replacement, 82-83

W

water spot, 98
web clamps, 35
white rings or spots, 98-100
  abrasive and lubricant use, 100
  removal, 98-100
wood chisel, 9
wood clamp, 11
wood cleaning, 109
  denatured alcohol, 109
wood joints, loose, 60-62
wood preparing, 110-111
  finishing, 110-111
  tack rag, 111
wood putty, 101-102
wood runners, table, 66-68
woodworker's glue, 13